MAKING CONSTRUCTION FUN AGAIN

AF722024

MAKING CONSTRUCTION FUN AGAIN

Structural Support for Those Building Our World

AMY POWELL

Copyright © 2025 by Amy Powell

All rights reserved. No part of this publication may be reproduced, stored in a retrieval system, or transmitted in any form or by any means—electronic, mechanical, photocopying, recording, or otherwise—without the prior written permission of the publisher, except for brief quotations used in reviews, articles, or educational settings.

Organizations and individuals are encouraged to use this book for team development, book clubs, or group training purposes. However, any reproduction, adaptation, or distribution of the material in part or in whole—especially for commercial, educational, or corporate use—requires prior written permission.

To request licensing, workshop facilitation, or educational use, please contact:
Well Works, LLC
www.livingwellworks.org
info@livingwellworks.org

This book is intended for informational and educational purposes only. It is not a substitute for professional advice, legal guidance, or mental health treatment. The author and publisher disclaim liability for any outcomes resulting from the use or interpretation of content within this book.

Well Works™ is a trademark of Well Works, LLC. All other product names, trademarks, and registered trademarks are property of their respective owners. Their inclusion does not imply endorsement or affiliation. We use names in an editorial fashion only, and to the benefit of the trademark owner, with no intention of infringement of the trademark.

ISBN 979-8-9925214-0-5 (paperback)
ISBN 979-8-9925214-2-9 (ebook)
ISBN 979-8-9925214-1-2 (audiobook)

Library of Congress Control Number: 2025911246

Cover and interior design by Cameron McKague

Printed in the United States of America

First Edition
10 9 8 7 6 5 4 3 2 1

To the field leaders whose wisdom built my foundation, whose support strengthened my structure, and whose belief reinforced my character—teaching me that building is about people, not just structures—and inspiring me to support and champion those who are building our world today.

Contents

Introduction

"I like things to happen, and if they don't happen, I like to make them happen."
WINSTON CHURCHILL

THROUGHOUT MY career in construction, I have repeatedly heard one phrase: "It's just not fun anymore." It's a sentiment that has echoed through jobsites, project trailers, and even casual industry conversations.

At first, I dismissed it as part of the job's natural ups-and-downs. I figured the discomforts and burdens of our industry were carried mainly by those of us in the bowels of construction. But over time, I realized that it isn't just those of us on the front lines feeling this way. Our partners across the broader construction ecosystem are struggling, too.

And it's not simply about difficult clients, long hours, or relentless project demands. It runs deeper than that. As I kept hearing it from seasoned superintendents, project managers, tradespeople, architects, and executives, I realized it isn't just a passing complaint. It's a call for action.

As a construction professional, my goal has always been to support and encourage those working in our industry, to reignite their love for their craft, and to provide tools and strategies to help them find fulfillment in their careers. That's why I wrote this book.

Construction is more than a career or a job. It's a lifestyle. It requires grit, endurance, and problem-solving in ways few other industries do. And while, yes, it can be hard, it doesn't have to be—and shouldn't be—a place where stress overshadows enjoyment.

This book is for anyone in construction who's ever felt worn down by the weight of it all. No matter your role—whether you're running crews, managing projects, solving design issues, or keeping the wheels turning behind the scenes—you are part of what keeps this industry standing. And while the work will never be easy, it shouldn't wear you out or leave you burned out.

Inside this book, you'll find practical tools to navigate stress, reconnect with the pride in your craft, and rediscover what made this work worth doing in the first place. We can't always control the challenges we face, but we can control how we view, respond to, and experience them. This is where the work of *making construction fun again* truly begins.

"It's Just Not Fun Anymore"

While I have heard this comment more times than I can count throughout my career in construction, there are three distinct instances where I recall hearing that construction wasn't fun anymore.

The first was around 2010, during a medical office building (MOB) remodel that I was managing. It was a challenging project that required us to work outside of the occupants' working hours, and it involved a demanding owner. Late one afternoon, a few months into the project, my superintendent and I were walking the site to check on progress from the night before and discuss any upcoming needs or challenges that needed to be tackled. During this walk, my superintendent confided in me that he had gotten into an argument with his wife. The argument had started because the project's difficult hours were putting a strain on their relationship. He wasn't whining, complaining, or trying to get out of doing his job; he was simply communicating a struggle and a challenge, colleague-to-colleague, friend-to-friend. Still uncertain as to how he could relieve some of this uncomfortable pressure, he lowered and shook his head. "This just isn't fun anymore," he said.

While I wanted to help, at that point in my career I wasn't sure how to. The best tool I had—and that I used at the time—was to be present and to listen to him. Looking back, I could have done more to help or at least asked him how I could help. Fortunately, it worked out on this occasion: he remained in the industry and is still happily married to his wife.

The second time I remember hearing this comment was around 2015, during a casual conversation in a jobsite trailer with a senior superintendent. We had just completed a training session focused on the impact of leading and running weekly subcontractor meetings. After the other field members had left the trailer and returned to their tasks, the senior super stayed in the trailer to chat. It was only a couple of minutes before the conversation organically shifted to his plan to retire within the next couple of years, finishing with, "Amy, this just isn't fun anymore."

Even though this was a number of years after the last time I'd heard this remark, I found myself yet again not knowing how to respond. My inability to offer solutions may have been the exact reason he'd chosen to tell me instead of his on-site crewmates. It felt as though he was relieved—like he'd finally gotten something off his chest that had been weighing on him for too long. This lack of enjoyment may have even been one of the strongest motivations behind his decision to retire. It was almost as if he didn't want to retire, but the enjoyable aspects of his work had slowly eroded over time. His perception of the job had turned stale and rotten, leaving a sour taste from a once-loved career. When he eventually hung up his safety vest and hard hat for the last time, I'm afraid it was with that sour taste about the industry and his career, rather than the sweet sense of pride, fulfillment, or accomplishment his years of hard work deserved.

I pour a lot into this industry. I don't want to reach the end of my career with a bitter taste in my mouth, feeling like none of it was worth it. I don't imagine you do either.

Up until this point, I hadn't really stopped long enough to consider how I felt about the industry or my own career trajectory. I'd dove headfirst into construction as a young adult and just kept swimming—without taking time to look at what I'd jumped into, what the challenges or benefits might be, or where I was even heading. It had seemed fun—energizing, even. But now, years later, I was starting to hear and see the deeper challenges I hadn't considered before.

Many of us find ourselves in this kind of situation: stuck in the weeds, so distracted by day-to-day tasks that we don't stop to consider where we're going or where we actually want to end up. After being confronted with this dilemma a second time, I realized I didn't want to just keep my head down and keep moving, only to wake up years later wondering how I'd gotten to a bitter-tasting retirement. I wanted to be more intentional—about my contributions, my impact, and the kind of mindset I'd carry with me when it was my turn to hang up the vest and hard hat.

So I paused to take inventory. I needed to ask myself: *Wait a minute. Am I on a path toward a bitter-tasting retirement? Or am I contributing to something meaningful, something fulfilling?*

I started exploring, mapping, and mentally tallying the pros and cons of working in the construction industry. The idea was to form a better picture of what success looked like to *me*—and whether I was actually headed in that direction.

What exactly had I jumped into? How deep was it? How far was it to the other side? What was on the other side? Who or what else was swimming against me—or with me? Should I turn back or keep swimming? What did I like about the industry? What did I despise? Had I jumped into the right career? Was this fun enough—and worth the effort—to keep swimming, or should I press the exit button and start over somewhere else?

The third and most recent time I heard someone express this concern came in November 2023, over a cup of coffee with a managing partner at an architectural firm. It was a timely conversation, as just a month earlier I had committed to writing this book. While I hadn't landed on a title yet, I was clear and committed to its purpose: to support and encourage those working in our industry, to reignite their love for the craft, and to provide relevant tools and resources to help them find fulfillment in construction and their careers.

Although the managing partner had phrased it differently than the two superintendents I'd spoken with years earlier, her comment reflected a similar, genuine grievance about the dwindling fun in our industry: *"Amy, I just want it [construction] to be fun again,"* she said. It put words to what I had been sensing for years, and our energetic conversation ultimately inspired the title of this book.

Clarifying and Redefining "Fun"

What would be your immediate reaction to getting cut by rebar sticking out of a slab on a jobsite? I would submit that it is not likely: "Eish, my shin is bleeding from a cut I got from that rebar over there that someone failed to cap! Can you take my tools, please, so I can bandage this and fill out an incident report?" It is much more likely that you'd shout some colorful expletives, or at the very least, "Ouch!"

In those first few moments, we may not know exactly what has happened, how it happened, or how deep or critical the cut is, but our body's pain sensors have already identified a wound and sent messages to our brain to alert those around us that we have been injured and may need help. Pain is our body's way of signaling that something is wrong and that we need to pay immediate attention to fix it. The initial profanity that follows an injury doesn't explain what happened, where it hurts, or what kind of help you need to get patched up and back on your feet. It also doesn't tell your crew to grab your tools, bring the first-aid kit, and cap that dang rebar! Profanity—or, "Ouch!"—is just our immediate response to pain, a call to action to those around us.

Hearing or saying, "Construction isn't fun anymore," is equivalent to the profanity spoken after getting cut from uncapped rebar. It's not about a lack of ball pits, beer, or BBQs. "Fun" is just the quickest and easiest word to articulate a deeper source of pain or discomfort to our packs—an attempt to alert others and recruit help. And while our brains are doing their best with the tools and language available to communicate that discomfort, many of us still struggle to put a finger on exactly what's causing the pain; how it's affecting our workmates, families, and the industry; and what it might take to fix it.

Since the problem here isn't a surface-level wound, accurately identifying and resolving it is much more challenging than bandaging a fresh cut on our leg.

In this book, I use the word "fun" to describe what is lacking and causing this feeling of unrest, since this is the term that's most often used in casual conversations on jobsites and among construction comrades willing to listen to our pleas. Fun in construction, then, isn't about eliminating challenges, but about how we approach them, and how we choose to respond to them. It's about building strong teams, fostering a culture of respect and learning, and finding real fulfillment in our work created by overcoming obstacles and tackling challenges together.

"Fun" in construction is that sense of pride, accomplishment, and camaraderie you earn from working relentlessly *with* your teams on very difficult yet rewarding projects. It's the fulfillment, achievement, and enjoyment you earn from perseverance, hard work, and fortitude year after year. It isn't necessarily about job or title promotions, but more about the self-assurance of your personal and professional growth, and the grit and team camaraderie that reward you in ways that material and monetary methods never could. (Although these are always nice as a cherry on top.)

"Fun" redefined: challenging yet enjoyable, fulfilling, and rewarding experiences.

As a project manager, my role was pretty straightforward: support our superintendents and field leaders, and make sure they had what they needed to keep their projects moving. Do that well, and things usually went well. But over time I started to see a pattern. The teams that truly thrived—where people enjoyed the work, where camaraderie was palpable, and work ran effortlessly and smoothly—had something else going for them. It wasn't just solid management or technical skills (although these are important, too). It was leadership. Emotional intelligence. Communication. The skills that don't show up on a blueprint and aren't lined out in an instruction manual or in the specs. But they undeniably show up on a jobsite.

And here's the thing: I spent years searching for ways to ignite these skills, and not just for myself, but for the people I worked with and cared about. The folks out in the field, leading the charge every day. I kept looking for support, tools, or resources that were actually relevant to our world in construction. Something applicable. Something that spoke our languages. But there just wasn't enough out there to help us with these "soft skills"—or, as I like to call them, "essential skills"—in a way that made sense and was worth our time.

I later realized that project or team success wasn't just about processes and systems. It was about people. It wasn't just the folks in the office who wanted this structural support and resources—our boots-on-the-ground leaders wanted it just as much, if not more.

Tired of waiting and searching for this critical support, I rolled up my sleeves. I went back to school and earned a Master's in Adult Education and Training (which was was no easy feat, as school never came easily to me). Then I set out to build what I'd been looking for all those years: a leadership development program designed specifically for field leaders. I chose them intentionally because in most companies, field leaders are the backbone of the operation, yet they're often overlooked when it comes to dedicated training and support. While office staff and management also often lack practical leadership tools, they typically have greater access to resources, mentorship, and development opportunities. Our folks in the field deserve the same kind of support—translated into experiences and language they can relate to and quickly apply to their daily work. This is how my company Well Works, and the Well Works Leadership Development Program, came to life.

Since then, the program has grown in ways I never expected. What started as something built specifically to support our field leaders has expanded to include workshops with craft professionals, executives, and even architects—while still providing direct coaching and support to those leading the work on-site every day. Being able to engage with people across all levels of the industry has given me a wider lens—one that extends from the field to the front office.

But no matter how far the program has reached, I've intentionally stayed close to the people I've always respected most: those out in the field, building the work with their hands, their minds, and their teams. That's where this started, and that's where my loyalty remains. They're the core of our industry, and when we support them well, the impact carries through the entire ecosystem, strengthening all of us.

This book isn't the program itself, but it's grounded in all I've learned from creating and delivering it: from the real conversations I've had with field leaders, executives, and peers around the country. It includes the hard lessons—the stories, mistakes, and breakthroughs that have helped translate these essential leadership skills into something practical that actually works out there, on real projects, with real teams.

The Wet Drywall Approach

When our work stops being enjoyable, it impacts every level of our industry. High turnover rates, disengagement, and declining craftsmanship are just a few of the symptoms when the human element is being neglected.

Ignoring fun and fulfillment doesn't just impact individuals either; it affects our entire construction ecosystem. Projects suffer when teams lack motivation, communication breaks down, and stress leads to costly mistakes.

Let's Take the Wet Drywall Approach

Imagine you are a contractor walking through the halls of your recently completed and occupied building. During a routine warranty walk, you notice a wall section showing signs of water damage. A rather large section of drywall is damp and water-stained near the base of the wall. Besides just putting it on the warranty list and sending it to the drywaller to fix it, do you:

A Cut into the wall, remove the damaged drywall, replace and patch it, and then slap on a fresh coat of paint (corner to corner, of course)?

or

B Investigate the cause of the leak creating the water damage, fix the leak, and *then* replace and patch the damaged drywall and paint?

or

C Do nothing and keep walking.

Working in the industry for over two decades, I am confident most of you would select "B": you'd dig into, find, and fix the leak causing the water damage first, and then replace the drywall. Sure, it requires a lot more time and energy, but it also effectively addresses and solves the problem.

Selecting option "A" creates a higher likelihood of having to replace the drywall again, because the leak itself hasn't been properly located, identified, and fixed. Your clients would be disappointed because they have invested in you to build a quality structure and trusted you to take pride and responsibility in your work. Continuing to have to fix the same problem time and again would indicate otherwise. You may also feel embarrassed and ashamed having to face the client and your boss to explain why you have had to return to address the same issue more than once. It's a ready-fire-aim approach.

Now, if this problem were related to essential (soft) skills, many of us naïvely or unintentionally select "A" and try to solve them using surface-level patches and quick fixes. Although this is done with good intentions, when we know more about the nature of the problem (the drywall), then we're more likely to tackle the deeper cause. If we know less about the subject (soft skills), then we're more likely to address only the symptoms.

Consider me the master plumber of essential skills. I've made it my business to tackle these hardest-to-teach essential skills and get them into the hands of our hardest-to-reach field members—those who are constantly on the move, working long hours in remote locations, and rarely tied to a desk. By focusing on the toughest challenges at both ends, the training becomes stronger, more practical, and more accessible for everyone across the board, no matter where they work or how they learn. I take pride in breaking these concepts down so they're easier to understand, are translated into the language of construction, and then packaged, delivered, and ready to be used immediately—on projects, with teams, and out on the jobsite.

The source (i.e., the defect or damage causing the leak) of this complex industry challenge of making construction fun again (i.e., replacing the drywall and pretty paint) is buried within and behind these essential skills. Unfortunately, this leak isn't water. It is the soft, squishy, uncomfortable realm of feelings and emotions. Now, before you feel the urge to close the book and go wash your hands, allow me to translate and explain.

In my years in construction and while developing leadership and communication programs for those in our industry, I had done everything I could to stay away from a few "touchy-feely" trigger words, including "feelings," "soft skills," and "emotions." Like many others in the industry, I was taught and told to "keep emotions out of it." I understood that to mean not being emotional, overly vulnerable, or dependent. I fully appreciate and understand the merit of this but for now, allow me to clarify and recalibrate our understanding of emotions and feelings the best I can without making you (or me) uncomfortable.

The challenge I faced was to decode and translate communication, conflict, efficient teamwork, effective leadership, and so on, while avoiding emotions and feelings. In my years of digging through research, however, I unexpectedly discovered that emotions and feelings are not only critical elements deeply embedded in the foundations of these essential construction skills, but they are also the roots of our logical decisions. Not only are they unavoidable, but they are the core of being human. It is less about keeping emotions out of it and more about navigating and understanding them, because they show up whether we want them to or not. If we were to remove these, we would be removing the human element.

Essential skills are critical to our industry—as important as knowing how to use and maintain your tools and how to navigate blueprints. Yet we are so opposed to emotions and feelings that we have intentionally tried to separate them, to keep them locked up at home and out of our construction processes. We have confused emotions and feelings for a lack of emotional intelligence and emotional control. This has caused us to unintentionally filter out emotions from construction altogether—but these are what motivate us as humans to produce physical structures and make logical decisions. This has not only caused frustration and unrest in trying to find that elusive "work-life balance," but it has also created challenges at work when we are told that we aren't showing up as self-motivated, loyal, dedicated, or engaged as we are supposed to be. For many years, we have tried to dissect, separate, and remove these critical elements, and by doing so have greatly misunderstood them and their value, mislabeling them as unproductive and useless in the workplace. When we separate and remove the elements that make us human, this is when we show up to work as "just another body," a zombie, or, more accurately, another piece of equipment. In our attempts to get more productivity out of our workforce, we've unintentionally engineered out the human elements of our construction processes—including any tolerance for human error.

But, here is the kicker: fun is a feeling. While one may still argue that fun doesn't fit into our stoic industry, respect and trust are essential to our industry aren't they? Both of these also happen to be feelings. We have unintentionally thrown the baby out with the bathwater, and the consequences are surfacing: isolation, self-preservation, turnover, stress, burnout, self-medicating, poor mental health, and high suicide rates.

The Plan

Let's approach this "fun" challenge like wet drywall. Before we start hacking away at the drywall with a saw and sledgehammer, damaging anything and everything behind it, we need a plan to ensure we aren't creating a more significant leak or creating a larger, more substantial problem.

Step 1: Acknowledging the Problem (Chapter 1)

Recognize and acknowledge that we do indeed have a problem or challenge: construction isn't fun anymore for many of our industry members, and we want to make it fun for ourselves and our comrades again.

Step 2: Locating the Problem (Chapters 2 and 3)

Construction not being fun anymore—the water-damaged drywall—is just a surface-level sign of a deeper problem in our industry, like the hidden cause of a leak. It means we need to dig in, observe, and find the root cause. We need to ask: what made construction less fun, and when did that shift happen? This pair of chapters begins by looking at the last 140 years of construction and the changes it has undergone, then turns to voices with longstanding industry experience to see if there was a point when the work stopped being fun for the majority of us.

Step 3: Assessing and Planning the Repair (Chapters 4 and 5)

Once the root cause has been identified, we need the proper tools, information, resources, and expertise to evaluate the possible solutions, barriers, and fixes—think as-builts, stud-finders, consulting the plumber and electrician. Identify what else is behind the wall that we may potentially damage if we don't properly plan. In these chapters, we figure out exactly what needs fixing, line out our tools we need to fix the problem, and prepare for any unforeseen surprises before tearing into that wall.

Step 4: Fixing the Problem (Chapter 6)

Once the source of the problem is identified and we have the proper planning and tools in place, we are equipped to solve it. From here, we start the remedial work by providing the best possible solutions based on what is known about the problem, the appropriate resources, and the information and tools available to tackle and fix the problem.

Moving Forward

This book did not come easily. It took energy, effort, and the hard work of confronting my own mistakes—and taking responsibility for them. More importantly, it took years of listening to colleagues, mentors, those struggling in the field, and those leading at the top. Construction is built on generations of lessons, resilience, and hard-earned wisdom. My goal is to honor everyone who has contributed to this industry, while offering support and encouragement to those still in it today.

This book isn't just about identifying problems. It's about finding solutions. It's about shifting our focus from simply surviving to truly thriving. By adjusting our mindset, building solid relationships with the people we work with, and embracing the inherent challenges and rewards of our industry, we can make construction fun again—not just for ourselves, but for future generations of builders.

In the chapters ahead, we'll see how our industry's history has shaped who we are and how we work today, dig into the deeper root cause of our "fun" problem, and, most importantly, uncover practical strategies to reclaim enjoyment in our everyday work. Along the way, we'll navigate the key balances, like getting the job done efficiently without sacrificing people, staying challenged without burning out, and honoring proven processes while embracing needed change. By blending hard and soft skills, we restore the sense of pride, camaraderie, and resilience that has always been at the heart of our industry.

So, let's get started.

STEP ONE

ACKNOWLEDGING THE PROBLEM

1

Why "Fun" Matters When You're Building the World

"The problem is not the problem. The problem is your attitude about the problem."

UNKNOWN

A FEW YEARS before I found myself at a career crossroads, I found myself in a hospital bed in the middle of the night. I had woken up around 11.30p.m. with terrible stomach pains, hyperventilating, and unable to take normal breaths. It got to the point where my hands clawed up and I couldn't get myself out of a fetal position. After a few hours in the hospital bed, hooked up to an IV and finally able to breathe normally, the doctor sent me home without a diagnosis, confident I would be just fine.

I was left to my own diagnosis, which I concluded was a buildup of weeks of unmanaged stress from a particularly challenging project that had finally taken me down. I hadn't been getting much sleep, and the sleep I *was* getting consisted of nightmares about crooked elevator shafts and delayed steel deliveries. I had been working long hours and pouring myself a drink almost every night after work as a poor attempt to relieve tension and quiet the constant thoughts running through my head.

At this point, stress wasn't just consuming me—it was digesting me. The pace of work, lack of sleep, lack of water, and an increase in mental and physical strain had caused my body to throw up a flag and shout (in its own language): "This isn't working, stupid! This is your only warning: either do something about it, or I will."

At the time, I assumed this was just how construction worked—and concluded that maybe I just wasn't cut out for it. But what if that wasn't really the case? What if stress isn't a badge of honor, but a warning sign? A symptom of a deeper cause?

I don't give up that easily. Call it stubbornness or fortitude, but I was determined to fight for my career. Nevertheless, the level of stress I was experiencing was clearly not sustainable. So I did what any self-respecting project manager would do: I set out to find the root of the problem—not only for myself, but also for my teammates. If I was going to reduce my stress levels and prevent another $1,500 visit to the emergency room for what was essentially a very expensive drink of water, then I needed to set aside my ego and invest some serious effort into solving the problem.

The trouble was, I didn't quite know how.

This is when I decided to roll up my sleeves and embark on a journey to identify the main causes of stress in construction, and how to respond to these stressors, either by increasing my tolerance to stress or reducing my stress response to a more sustainable level—not just for myself, but for my teammates, colleagues, and industry brothers and sisters. While this journey is ongoing, I've uncovered valuable insights, tools, and potential solutions that can help us reduce stress while increasing fun, fulfillment, and enjoyment in our line of work.

Blaming, finger-pointing, siloing, dependency, denial, and avoidance don't work. In fact, they only contribute to creating bigger problems. We all know this, even if admitting it is hard. Facing these issues head-on is challenging, but, as you'll read in this book, it's also necessary.

So let's talk about fun in construction—the importance of it, and how to go about efficiently and effectively igniting it (again) and maintaining it.

Problems and Stress

What makes the construction industry fun?

What makes it unbearable?

What makes you either stick with it and keep swimming or decide to quit?

When I was considering what I enjoyed about the construction industry, what made it fun, three reasons stood out above the rest that I have since found resonate with many others as well:

The challenge and momentum. No two projects, teams, or weeks are the same. This provides momentum and endless opportunities to keep busy. Different challenges, people, and environments allow us to avoid stagnation and boredom.

The people and camaraderie. The types of people our industry typically attracts are my favorite people to be around. They're the grittiest, most enterprising, loyal, and authentic people I know, and I enjoy working with them, learning from them, and representing them. Sure, they can be a pain in the butt at times, but they're the main reason I love this industry so much.

The opportunity to contribute to something impactful and important. Something essential, even critical to our communities and society; something much larger than ourselves. A legacy.

This realization and experience in hunting and solving these problems for a number of years hit me like a ton of bricks: The "problems" weren't the problem. It was my *mindset about these problems* that was the problem.

While problems were on my industry hit list for years, it was at this point—while reflecting on the things I enjoyed about the industry, sitting on a proverbial career-break island, mapping out and contemplating my career trajectory—that I realized I had "problems" listed in the wrong column.

Some problems spark creativity and team pride, but unrelenting or poorly managed problems can lead to burnout rather than motivation. So striking a balance—fostering the challenges we can handle with our team without being overwhelmed—is key to making the problem-solving process fulfilling, effective, and sustainable.

If we were to somehow eliminate problems altogether, there would be nothing for any of us to come together for, to overcome, which would evaporate any chance or opportunity for camaraderie. The very things I love about the construction industry rely on the existence of problems. At this point, I mentally erased "problems" from the "industry cons" list and added them to the "pros" list.

While this exercise did not eliminate my stress, it did reframe what stress was and could be, and increased my tolerance, which made working in construction more fun. I stopped drinking every night. I started seeing problems not as obstacles preventing me from moving forward, but as exciting opportunities and adventures to overcome, work around, and figure out. I came to realize that while I can do my best to try and prevent as many problems as possible, there will always be those that cannot be prevented or planned for. Their only power is what we decide to give them.

Our power lies in our mindset and perspective. Are problems a barrier, or are they an opportunity to learn, grow, and bring together fantastic people who can combine their strengths and skills to overcome obstacles? The number of problems—and if or when they arise on our projects—is outside our control. But how we choose to label, view, and respond to those problems is entirely within our control.

Problems are what make our industry exciting and challenging. If we were to eliminate all problems, we would have nothing to solve. This would promptly create monotony, boredom, and stagnation.

Reintroduction to Construction

Before we dive into fun, allow me to reintroduce you to construction. At first, this might seem like an unnecessary step in a book written for those already in the industry. However, if you've ever tried to explain to someone outside our field who we are, what we do, and how and why we do it—the intelligence, planning, effort, and collaboration it requires—you'll understand why this refresher is essential before we dive in.

Construction is amazing. We are the heartbeat of America and the communities we work in. Others cannot do most of what they do without us doing what we do first. We build and maintain structures where people live, gather, celebrate, heal, learn, and work. We build and maintain infrastructure that provides these places with clean water, reliable power, HVAC, and disposal of sanitary waste. We build and maintain roads, bridges, airports, and sidewalks to safely transport ourselves and our loved ones to these social lifelines.

In her book *I, Mammal*, Loretta Graziano Breuning does a fantastic job of painting a picture of what it would be like if we didn't show up every day and do what we do:

> If you were born before flush toilets and electricity, you would have gone to the vermin-infested outhouse in the dark. Cesspools and kerosene lamps would have filled your world with foul odors, toxins, and disease. Animals would have tracked dirt between your outhouse and your cooking area, creating a vector for intestinal worms that would steal the nutrition from the food inside your body.[1]

Those of us in construction are the people who show up every day, project after project, building the foundations society depends on—because the world can't do all that it does unless we do what we do first.

While our industry is indeed amazing, and the things we create are amazing, it is also exceedingly complex and

---

1 Loretta Graziano Breuning, *I, Mammal: How to Make Peace With the Animal Urge for Social Power* (Hayward: Inner Mammal Institute, 2011), 221.

challenging. A common but detrimental assumption is to think that these challenges subside over time or with career advancement. That leadership is some type of destination that we earn our way to—a sandy beach where we can wear flip-flops and have our own wait staff. The reality is far from a sandy beach, and the challenges frequently intensify over time, with advancement and experience. Responsibilities increase. You're not just responsible for your tools and daily checklist, but the lives of the other people on site and the success of the entire project.

This means that remaining within the industry requires strength, fortitude, and gumption. When we build this endurance, we can more easily identify and overcome more complex problems. While this type of environment may seem like hell to some, it attracts and retains some of the best people and provides the most rewarding experiences and opportunities to build onto your own legacy.

In the first few years of my career, I recall many experienced and wise industry members offering genuine advice that construction is an industry that requires hard work and thick skin. Naïvely, like many of us in our youth, I disregarded this advice, thinking I already knew how difficult it would be, and arrogantly believed I was already juggling and handling these so-called difficult tasks just fine. I had finished busting my hump in school to learn how to manage projects, had earned some respect from my teammates, had stuffed myself into the standard-issued stoic bodysuit of thick skin, and was already putting in long hours. On top of this, I was nailing the whole "work-life balance" thing—being able to go out and drink at night, come to work mostly attentive the next day, and still feeling I was able to run circles around those old farts in the field.

Looking back at myself, after many, many years of living, breathing, sweating, crying, and bleeding this industry, I realize that no number of words could have effectively described how hard construction is. It's difficult to imagine something you haven't yet fully experienced. Even with experience, we often think of only the wear and tear on our bodies. The early-morning concrete pours; off-hour HVAC swaps; the long, grueling hours of the "summer wonders"; six-tens; undocumented overtime; the freezing cold and unrelenting heat; and the awkward, yoga-like positions required to squeeze above acoustical tiles or behind walls—all of which most definitely take a physical toll.

Yet, along with all the physical exhaustion we endure, there's another side to this job—one that's often tougher to handle in the long term: the mental toll. It's more than just managing schedules or solving daily problems on-site. It's the weight of making decisions that don't always have clear answers. It's knowing that a single misstep—in planning, execution, or oversight—can have real consequences. It's the constant pressure of feeling responsible not just for getting the job done, but for making sure everyone who makes it happen stays safe and healthy. While we might think we can handle it all, sometimes reality forces us to stop and truly recognize the weight we carry—and how important it is to balance it.

Our Responsibility

It's incredible how words take on new meaning as we go through life. Like water slowly carving a deeper canyon over time, experience reshapes these words and gives them more weight. Take "responsibility," for example. At first, I thought responsibility was something you earned by climbing the ranks. I realize now I had confused it with authority.

Early in my career, I saw responsibility as measured in deliverables—deadlines met, budgets balanced, productivity maintained. If a project stayed on schedule and within budget, I assumed I had done my job well and met my responsibilities. In fact, I was nailing the technical aspect of responsibility without yet fully grasping the human side of it.

The meaning and weight of responsibility truly hit home for me one day after a tragedy on another worksite. As I was leaving the jobsite of a commercial project I was managing, I overheard worried murmurs from my team about a trench collapse at a nearby residential site. When I got home, I turned on the news. Nearly sixty rescuers had been working tirelessly for hours, trying to save a young construction worker trapped under fifteen feet of dirt. His teammate had already lost his life in the collapse, buried somewhere in the same trench.

The rescuers moved as fast as they could, digging by hand, using buckets and shovels. Using an excavator was out of the question—the displaced soil would have crushed the trapped worker instantly. For hours after the collapse, he was still alive. Still breathing. Still talking to the rescuers through a site utility pipe—the very pipe he had likely installed earlier that day. They contacted his family, who were able to talk with him through that same pipe one last time. And then—silence.

The rescuers couldn't reach him in time. He didn't make it.

I sat there, feeling the weight of the moment sink deep into my gut. I'm not much of a crier. But it stayed with me. I found myself thinking about it—and crying about it—for weeks afterward. Like that slow carving of a canyon I mentioned earlier, it just settled in a little deeper each day.

And beneath the grief, there was anger. This accident could have been avoided completely if the proper safety measures had been taken. I thought about the project manager and the superintendent on that job—the weight they must have carried, knowing two men had died under their watch. I imagined what it would feel like to hear my husband's last words through a PVC pipe. I thought about the man's crewmates, the ones who had worked beside him that morning, who had seen the collapse, who likely started digging with their hands or whatever tools they could find to uncover their teammate. The ones who ran for help but in the end could do nothing else but wait. I thought about the families, their grief, their loss.

Two lives lost. Two families forever changed. An industry forever scarred. The impact that rippled far beyond that jobsite.

In construction, tragedies like this are, unfortunately, all too common. Sometimes we even live through them. When we do, we carry them with us. It feels like a punch to the gut, except, unlike a physical wound, this one doesn't heal. This industry demands a lot from us—the mental strength, the endurance, the weight of responsibility. It can be overwhelming, but it is also one of the most meaningful, rewarding careers imaginable.

This tragedy taught an impactful lesson; responsibility took on a new meaning. It's not simply about budgets and deadlines. It's about people.

We don't just build structures. We build opportunities. We build environments where people can thrive, not just survive. When responsibility is taken seriously, we create workplaces where people don't just endure pressure—they grow under it. When the full scope of responsibility is ignored, though, corners get cut, safety is overlooked, and worst of all, people get hurt. This is the real weight of our work. It's why responsibility, when fully embraced, isn't a burden. It's a privilege.

Our work matters. The people in it matter. If we learn to own and carry responsibility with purpose, it doesn't just make this industry rewarding—it shapes us into leaders who have lasting impacts.

So we build strength and endurance to carry this weight. Learning from the hard moments. We take responsibility not because it's easy, but because it matters. And if we do it right—if we lead with both strength and purpose—we don't just build projects. We build legacies.

To the young men who lost their lives that day, and to their families who have carried the weight of this loss: I carry you in my thoughts. Your stories are not forgotten. You remind us why safety, responsibility, and looking out for each other can never come second. May your legacy push us all to do better, so no one else has to face a tragedy like this again.

Choose Your Hard

I fully realize that not everyone can—or wants to—move problems from their cons list to their pros list. Some folks have a cons list so long that it threatens to tip the whole dang scale. Honestly, if I thought the rest of my career was going to be miserable, I might've started looking for the exit, too.

But walking away isn't exactly a walk in the park either. If you're near retirement—like my senior super, introduced earlier—you might be close enough to a warm beach to call it. You could bushwhack your way onto a new path, or, heck, build a lakeside cabin and spend your days on the porch yelling at pesky kids. But most of us are somewhere between the beginning and the middle of the trail—mud on our boots, pack on our backs, and a long way from any exit—where walking away just because the job "isn't fun anymore" isn't exactly an option.

Like me, you've probably got a lot invested in this industry, too, which makes it hard to cut your losses and start over somewhere else. I'd collected wisdom from seasoned vets, earned a four-year degree, a three-year master's degree, and a laundry list of certs—SWMP, OSHA, LEED, ASHE, WELL, Ramset, Genie Lift Operator. I've got loads of failures and wins, hard lessons and proud moments—but more than anything, I've invested time and heart into relationships with others throughout the industry.

Starting from scratch isn't simple. What if another career has just as many bad days, or more of them, and worse ones at that? Clacking away at a keyboard for seven hours might sound easier, but would it be more fulfilling? Or retail, where you can just show up, check off someone else's list, and go home at the end of the day and sleep without being haunted by work nightmares?

Being handed a pre-built, prioritized checklist sounds refreshing at times. Especially on those days when you don't even get a chance to *look* at your to-do list. Days when you dread walking into the trailer because it will be there, sitting, waiting for you, like your mom that time you completely blew off curfew. Or the days when adulting is required to not only create our own to-do list but prioritize a heaping pile of tasks that surpass the color options in a Sherwin Williams fan deck. It might be nice for a week or two, but being handed a to-do list all the time would become boring and stagnant for those of us who require more advanced challenges that require more brain power.

So, what are we going to do about it?

STEP TWO

LOCATING THE PROBLEM

2

A Field Trip Through Construction's Past

"The longer you can look back,
the farther you can look forward."

WINSTON CHURCHILL

CONSTRUCTION NOT BEING fun anymore (the water-damaged drywall) is the surface-level symptom of a deeper problem or challenge (the cause of the leak) within our industry.

We've effectively acknowledged that we do indeed have a problem: many of our workmates are expressing their lack of fun, fulfillment, and engagement. So let's do our stretch and flex, throw on some gloves and safety glasses, grab a flashlight, and see if we can figure out where this leak is coming from!

What's clear is that the two words "anymore" and "again" are welded to every grievance about our industry not being fun.

The way these words show up so consistently has made me wonder about two things:

1. When did construction stop being fun?
2. Was there a specific factor that caused our work to become less enjoyable?

In other words, is the lack of fun a new problem or something rooted deep in our industry? When did people last walk onto jobsites with a bounce in their step? And when did we start graying or balding by our mid-thirties and settle into our stereotypical cranky disposition?

The problems we face in construction aren't the kind that come in tidy little packages, wrapped in shiny paper with a bow. Even if they were, that's not really how we like our problems delivered. Our problems are gnarly, multifaceted challenges that take a crowbar, a survival pack, a first-aid kit, and a solid dose of critical thinking to unbox and solve. These are the kinds of problems that we have to hunt for, find, and solve before they find us. Thankfully, our knack for solving tough problems gives us a real edge in tackling our current fun conundrum.

To do that, we need to understand where we've come from. Looking back at the conditions, culture, and expectations of the past can help us understand what changed—and why. What parts of construction used to feel rewarding? When did that shift? What was lost along the way? To start answering these questions, let's travel back to the late 1800s and step into the boots of one of the earliest field reporters of construction work.

Using our top-notch construction problem solving skills, we're going to look back over 130 years in our industry to see if there are any trends to spot in the historical record.

Revisiting our past is critical. While it might be trendy to "LITM" (live in the moment), who we are today is built on a robust foundation of history. Tools, tech, and conditions may have changed, but our industry's DNA hasn't. It is still made up of people—and it's this that provides us with clues as to how we can build fun into our industry again.

So, let's take a jaunt through construction's past, from the 1890s to today.

A Week in the Life of a Day Laborer in 1891

The oldest, first-hand field account I could get my hands on came from an antique store in Wyoming. Dated 1897, *The Workers: An Experiment in Reality* by Walter Augustus Wyckoff captures a unique snapshot in time. Wyckoff, a Princeton grad, traveled across the US from 1891 to 1893 to gain an understanding of the lives of the working class. He admitted to being "vastly ignorant to the labor problem and trying to learn by experience."[2]

Despite huge differences in technology, the world Walter worked in isn't too far removed from today's jobsite. And though he worked in construction for just a week, his ability to describe what it felt like—rather than just critique from the sidelines—paints a powerful picture.

Let's put ourselves in Walter's shoes. First, erase the modern gear: phones, computers, GPS tools, vehicles, lunchboxes, boots, microwaves, and Sharpies. Your tape measure isn't self-retracting. Your underwear feels like forty-grit sandpaper. Laundry is a two-day job that can only be done when you

2 Walter A. Wyckoff, *The Workers: An Experiment in Reality* (New York: Charles Scribner's Sons, 1897), 2.

have that sort of time to spare. Lunch is a slice of pie and a sandwich wrapped in old newspaper.

The laborers walk to work in a group, along a road of gravel and dust. The car barely exists. You walk a mile alongside other laborers—silent, sweating, coats over arms, lunch pails in hand. Wyckoff captured that morning walk vividly:

> *Gangs of workingmen, as I recalled them, were uniform effects in earth-stained jeans and rugged countenances, rough with a varying growth of stubborn beard... It is a mile from Highland Falls to West Point, and we moved briskly. There was little conversation among the men. Most of them had taken off their coats, and with these over their arms and their dinner pails in hand, they walked in silence, with their eyes on the road.*[3]

Good thing you did your morning duty back at the boarding house, because at the jobsite there aren't, what would be considered today, any sanitary or private Port-O-Lets with the latest jobsite gossip artistically expressed on their walls. Instead, there is a privy shared by everyone on the jobsite and those living on the block you are working on: a wooden bench with holes cut out, balanced over a brick pit.

When you get to site, you don't have the authority to look at any blueprints to get an idea of what you are building. Just as well, as you likely haven't been taught how to read them anyway. There isn't any formal training on what to do, how to do it, or let alone how to do it safely. Your immediate task is shouted at you by "the boss," who is standing high on his post. Any part of his commands that are unclear will have to be worked out by observing what everyone else is doing and absorbing cues and commentary from your workmates.

---

3 Wyckoff, *The Workers*, 38–39.

A pile of rubble consisting of the remains of a former structure lies in heaps in front of you. A few days prior, dynamite was used to crumble it into its current form. The brick and plaster are too clunky and jagged to use shovels. So, using your bare hands, you remove the pieces of brick and plaster from the pile of rubble and toss them into carts, which when full are hauled off and returned by young men. Boys as young as ten years old haul water. Wyckoff summed up the grind like this:

There were simply ages of physical torture, and short intervals when the physical reaction was an ecstasy.[4]

You feel like "just a number," and that's because you are: you've been handed a brass disk with the number seven stamped on it by someone called the "timekeeper." You are instructed to wear it and show it every time they make their rounds, so they can keep track of the hours you work.

During your allocated breaks and between your laborious tasks, you and your workmates still talk with one another, using unique, instinctual, often non-verbal construction language. The sense of camaraderie is strong. Among the crews, English, Spanish, German, Dutch, and French can be heard. And a shared language of profanity binds them together. Even

4 Wyckoff, *The Workers*, 37.

though Walter is a college graduate and may come from a different societal class, he is still able to recognize and appreciate this:

They swear with the readiness of instinct... not merely in anger... but in commonest intercourse... as meaningless as casual interjections.[5]

Other than the allotted forty-five-minute break for lunch—spent scarfing down smooshed sandwiches and apple pie—you're sorting rubble for eight and a half grueling hours. Your hands are cut and bleeding. Your back and arms ache from bending, twisting, and contorting all day. With no ventilated hard hat, your head pounds from the sun beating down on you—and from the boss micromanaging your every move, barking orders if you pause outside the strict break schedule.

After just a week in the field, Wyckoff captured not only the physical toll, but the emotional and spiritual cost of labor under poor leadership. He summed it up in a quote that still lands today:

> All truly productive labor performed under right conditions is itself a blessing... It is a bringing of order out of chaos, a victory over forces which can be reduced from evil mastery to useful service... In this sense labor is prayer. But

---

5 Wyckoff, *The Workers*, 58.

> robbed of its true conditions and reduced under the eye of a slave-driving boss, who impels us with insults infinitely more degrading than the lash, labor is no longer prayer, but a blasphemy.[6]

While this quote may be over a hundred years old, it hits on an important point: when the physical aspects of our work are led with purpose and respect, it can be meaningful—even fulfilling. But when leadership strips away the dignity—through micromanagement, lack of support, or relentless pressure—the purpose drains out, and the pride goes with it. This might not be the whole answer to why construction stopped being fun, but it's definitely a piece of our puzzle.

The sun starts to set, and the head boss calls out, "Time's up!"

Yet, you don't get to sit down, enjoy a cushy car seat, listen to your favorite music or podcast, or feel the cool, refreshing blast of your vehicle's air-conditioning across your face. Just another mile-long trudge back to the boarding house you're staying at: you're working too far from your home and family to commute. Once inside, you clean off your dusty, bloody, swollen hands, eat a shared meal with four other boarders, and collapse upstairs in an unheated, un-air-conditioned room—shared with those same workmates. Your body throbs, your head still pounds, and you're exhausted. Then, sleep. Just enough to be able to drag yourself out of bed and do it all again tomorrow.

Women weren't yet part of this workforce. The only women were adjacent to this work: packing lunches or managing the boarding house or, if well-to-do, passing by on horse carts, dressed in fancy dresses. In Walter's era, the construction

---

6 Wyckoff, *The Workers*, 71–72.

industry is exclusively male, meaning it missed out on the insight and impact women would later bring to the field. It wouldn't be until the 1910s that women made an appearance on jobsites, in the role of supporting with admin and final cleaning.[7]

Moving Along to the 1920s to 1930s

Let's jump ahead thirty years or so, to the '20s and '30s. This was not an easy era for construction or for the world in general. In America, with Prohibition (1920–1933) and the Great Depression (1929–1939) in full swing, it was slim pickings when it came to digging up any fun or enjoyment. Finding a job was nearly impossible, and if you were lucky enough to land one, you'd do just about anything to keep it: working insane hours, taking unsafe risks, masking injuries, ignoring health issues—anything to avoid joining the ranks of the unemployed.

However, a few notable changes were making it easier to do your job more efficiently, tilting the balance slightly toward "fun" and away from "treacherous." The Thermos (1904) allowed you to bring hot coffee or soup to work. Washing machines (1907) meant your clothes could be cleaned more quickly. Band-Aids (1920) upgraded first-aid kits. Taylorism provided the earliest forms of project management.[8]

---

7 Frank G. Carpenter, *How the World is Housed* (New York: American Book Company, 1911).

8 Dr. Sherif Hashem, *Greatness in Construction History: Human Stories of Great People and Great Projects* (New York: Business Expert Press, 2022), 10.

Other game-changers like pre-stressed concrete (1928), the supermarket (1930), button-up shirts (1932), and ballpoint pens (1938) were just beginning to shape the way you lived and worked.[9]

Jumping to mid-1929, the industry is starting to design and build something called "skyscrapers." Three projects are currently competing in what's being called "The Race into the Sky," each aiming to become the tallest building in the world: the Bank of Manhattan Building, the Chrysler Building, and the Empire State Building. You are confident you could get work at the Empire State Building, which is expected to win the race. You decide to pack up your tools, books, and personal items, kiss the family goodbye for several months, and head to New York City to see if you can get a spot on the project team.

You're settled at a boarding house with other potential workmates when Tuesday, October 29 (Black Tuesday) rolls in like a heavy fog, choking out your sense of stability. That day, more than sixteen million shares were traded on the New York Stock Exchange, and the market crashed. In the months that followed, the crash—along with slowing demand, massive consumer debt, and decreased industrial production—triggered the worst economic downturn in modern history: the Great Depression.[10]

Sherif Hashem explains in *Greatness in Construction History*, "Luckily unlike most real estate developers, John Raskob,

---

9 Roger Bridgman, *1,000 Inventions and Discoveries* (New York: DK Children, 2020).

10 HISTORY.com Editors, "Great Depression History," last modified October 20, 2023, https://www.history.com/articles/great-depression-history.

Empire State Building project Investor and Owner, happened to have stopped speculating months before the stock market crash. That gave him a favorable credit status, enabling him to get the funding required to build the project."[11] This means your targeted project is still on track to start construction, while most other projects are halted, postponed, or abandoned. Nonetheless, there are delays while funding is confirmed.

The amount of pressure and stress to secure a role on this project and continue to provide for your family builds to a crushing weight. You can't have a drink to take the edge off. Remember, we are still in the era of prohibition, and alcohol would likely just add to your problems more than solve your predicament anyway. The project team isn't hiring just yet, but sitting around waiting and worrying isn't going to do you any good.

To fill the time, you turn to your twelve-volume *Radford's Cyclopedia of Building Construction*, which you bought after seeing it advertised in *American Builder* and had sent to your new address. The four-page ad featured a genie-like figure hovering above a well-dressed worker studying at a desk, accompanied by the bold promise: "Knowledge brings success and prosperity." The last page of the ad shows young workers entering a building made of books labeled "Success," while one laborer in overalls, tired, hunched over, and worn down, is shown turning away from the line. He's labeled "The Man Who Stood Still."

The message is clear: those who invest in learning and growth keep moving forward, while those who think they already know enough and stop investing in themselves

11 Hashem, *Greatness in Construction History*, 41.

eventually fall behind.[12] That $37.50 you spent on the twelve-volume set? It's already proving to be one of the smartest investments you've made, especially now that desperate workers are flooding into the city, all scrambling for limited work.

December 1929: funding for the Empire State Building is officially secured, and the project is awarded to Starrett Brothers and Eken; they are now officially hiring workers, and work can begin. On January 12, 1930, American Bridge is awarded the structural steel contract, and you are among the ironworkers given a spot on the crew.

There is one caveat: the client not only wants this to be the tallest building in the world, but also built in record-breaking time. This is not because potential tenants are knocking down the door wanting to occupy the building as soon as it's complete; the client simply wants to prove that it can be done. This means that 102 floors—2,248,355 square feet—need to be built and completed in 464 days. To accomplish this, you will be working a twelve-hour shift, along with anywhere between 300 and 3,400 other workers on-site. You get to pee in a communal bucket placed on whichever floor you are working on that day. You also get to take your cigarette and lunch breaks up there, because hoisting you up and down the floors of the building takes too much time, and time is money, right? If any of this start to feel overwhelming or exhausting, you'd better just suck it up, otherwise you risk losing a coveted spot on the crew and quickly rejoin the army of the unemployed.

On a positive note, you and your project teammates'

---

12 Radford's Cyclopedia of Building Construction, advertisement, *The American Builder*, January 1928, 53–56.

heroic and risky line of work is finally being noticed by the media, who have decided to write an article on the ironworkers working on the project. Countless photos are taken of you and your crew. Some teammates show off a bit too much—someone should remind them there's still a production quota to hit—but their existence finally being noticed, recognized, and appreciated by society seems to renew everyone's sense of purpose, creating an uplifting energy that buzzes throughout the entire jobsite. The article and photos will later reflect the uplifted spirits of the project team. Many smile and show off their skill and pride, feeling like the respected and productive members of society they truly are but are rarely recognized as being. The photographer descends the building's skeleton and heads back home to his family, while you and your crews continue working until there is no more daylight to work productively.

You have become one of the "Sky Boys," as one journalist describes you, working tirelessly, and at dangerous heights, to complete this project. Later, in 1932, another photo is taken of your crew on the Rockefeller Plaza project, while you are sitting and eating lunch on a recently placed beam 850 feet AFF (above finished floor). It becomes one of the most iconic photos in construction—and labor—history.

1940s Through the 1970s

Let's move along another generation or so. Microwaves and Tupperware arrived around 1946, letting you quickly reheat food at home and pack a decent lunch for work. Over the next few decades, technology continued to accelerate. Industrial robots sped up material production, push-button phones improved communication, and by the 1970s, floppy disks, calculators, personal computers, and early cell phones hinted at a digital future. Cameras helped document safety violations by the mid-1950s, and spreadsheet software in 1979 revolutionized how we tracked projects. The Occupational Safety and Health Administration (OSHA) arrived on the scene in 1971, ensuring that safety was no longer optional. And in 1969, as Lloyd Christmas would say, "WE LANDED ON THE MOON!"[13]

An insightful speech given by President John F. Kennedy at Rice University Stadium in Houston, Texas, on September 12, 1962, reminds and motivates us about the importance of tackling tough tasks. Something that can fuel our sense of fun and fulfillment within the industry:

> We choose to go to the Moon. We choose to go to the Moon, we choose to go to the Moon in this decade and do the other things, not because they are easy, but because they are hard; because that goal will serve to organize and measure the best of our energies and skills, because that challenge is one that we are willing to accept, one we are unwilling to postpone, and one which we intend to win, and the others, too.[14]

Alongside all the impactful inventions and tools of this era came a major growth spurt for the different sectors of our industry. World War II started September 1, 1939, which required many of the men in our industry to refocus their

13 *Dumb and Dumber*, directed by Bobby Farrelly and Peter Farrelly (1994; Los Angeles, CA: New Line Home Video, 1994), DVD.

14 Alan Axelrod, *100 Turning Points in American History* (Lyons: Lyons Press, 2019), 299.

energy from building our country to protecting it. Manufacturing pivoted to support military infrastructure—producing equipment, roads, and temporary structures. But while projects back home slowed, the demand didn't disappear. And this is where the women stepped in.

Rosie the Riveter became the face of a movement. Between 1940 and 1945, the percentage of women in the workforce jumped from twenty-seven percent to nearly thirty-seven percent. By 1945, nearly one in four married women were working outside the home.[15] In August 1943, *LIFE* magazine ran a feature on women working in steel mills—welding, running cranes, operating machinery—at the same pay rate as men.

> To present the first comprehensive picture of women toiling in heavy industry, *LIFE* sent Photographer Margaret Bourke-White to Gary, Ind., where 4,800 women are now employed in the plants of United States Steel Corporation subsidiaries... The women who have taken over men's jobs in steel are receiving the same rate of pay as the men. They work eight hours a day, six days a week... The length of this list is especially significant because before Pearl Harbor virtually the only women employed in the Gary steel mill area were the sorters in the tin mills.[16]

That article—and Rosie the Riveter—were game changers. It proved that hard work and skill had nothing to do with gender. Personally, I'm forever grateful to these women. They

15 HISTORY.com Editors, "Rosie the Riveter," last modified February 20, 2024, https://www.history.com/articles/rosie-the-riveter.

16 LIFE Magazine, "Women in Steel: They Are handling Tough Jobs in Heavy Industry," *LIFE*, August 1943, 75.

didn't just fill in temporarily; they opened doors. Thanks to them, I get to be part of this industry today and so do countless other women. I like to think I would've been right there beside them, welding in a steel mill.

After the war ended on September 2, 1945, all of the military research and development needed a new outlet. That's when our tools and equipment took a massive leap forward—bulldozers, power tools, and new methods that reduced jobs from weeks to days.

From 1946 to the mid-1960s, the government and private developers poured money into residential and commercial construction, turning fields into suburbs almost overnight. This building boom changed the face of America. Residential neighborhoods sprang up quickly, and commercial development quickly followed, bringing offices, schools, and shopping centers closer to home.

Meanwhile, the launch of the Interstate Highway System gave rise to a whole new sector of the industry: heavy civil. Highways transformed how we traveled and served as a contingency plan for moving military equipment in a time of crisis. Bridges, roads, and tunnels became as critical to national defense as to daily life.

If you were working in construction during this era, you'd have seen everything start to speed up. Bigger machines. New tools. Tighter schedules. The job might still demand blood, sweat, and grit, but now power tools replaced hand tools, and trucks were hauling in hours what used to take a week of labor to move. You'd likely be working alongside specialized crews, getting used to project managers who kept the chaos moving in some kind of order.

My friend Chris Harper—once a plumber, now a PhD construction management professor—helped me connect the dots. According to Chris, this post-war era was a massive turning point. The tools were better. The jobs were faster. And our entire approach started to shift. By the late 1960s, specialization was the name of the game. Trades like electrical, plumbing, and steel had their own crews. General contractors stopped trying to do it all, and instead started managing the specialists. As projects got more complex, owners brought in construction managers to keep timelines, budgets, and risks in check.

Then came the 1970s. OSHA was established with a new message: everyone should go home the same way they showed up. Environmental protections followed—encouraging us to think not just about what we build, but how we build it and what it does to the world around us. Stormwater, erosion, and pollution all had to be accounted for.

Chris summed it up best: this era didn't just shift how we build—it transformed who we are as an industry. Bigger machines, stronger safety standards, and a broader perspective on how we impact the world. But at our core, we were still the same: a crew of gritty, skilled problem-solvers just trying to build something that lasts.

Picking Back Up, 1980s–1990s

Moving right along to our next generation of workers, let's excavate a bit further into this era and see what was going on in our industry and the world at that time.

Like a VHS tape on high speed, we'll need to fast-forward through the sheer number of inventions. By the mid-'80s, we had the IBM PC, AutoCAD, SureTrak, and even a dust-free builder's chute on the jobsite. Windows software systems hit our computer screens, CDs replaced cassette tapes, and digital tools started pushing out hand-drawn plans. Then came the 1990s: the World Wide Web (text-only at first), digital cell phones, email, and the first graphical web browsers.[17]

These are indeed a big couple of decades for our industry. While some found the pace of innovation exciting and energizing, others found it downright overwhelming. We had larger equipment, better workwear, and more advanced personal protective equipment (PPE). We could even swing by gas stations or fast-food joints for hot coffee and meals delivered in disposable packaging. But our biggest upgrades came in the form of electronics.

You're still on the jobsite, but now you've got a pager on your belt. You've rolled up and stored your hand-drawn blueprints because now the plans are being printed from electronic files. You're starting to hear about something called "email," though most of the team still sends notes via inter-office envelopes—the kind with the red string and multiple crossed-off names.

The landline telephone is still your main lifeline on site, bolted to a post and shared by everyone. Although cell phones are beginning to pop up, they're far from universal. You still have to be intentional about being near a landline to take or make scheduled calls to team members. This means you have to be more prepared for discussions because you only have a short amount of time to work out any change orders, design changes, or discrepancies during the scheduled call time.

To make up for the communication lag, you start holding recurring, in-person project meetings. When the team's spread out, you use conference calls—thanks to new features

---

17 Bridgman, *1,000 Inventions and Discoveries.*

added to phones in the 1980s. All of this makes you more efficient, even when juggling multiple projects in different regions.

But the pace? It's picking up fast.

The '80s and '90s seemed to throw in a whole new set of "fun hurdles" for our industry to navigate. General contractors (GCs) were getting a bit scrappy and bossy, operating with a feast-or-famine mindset in the early 1990s—pushing more risk and pressure onto our fellow subcontractors. Additionally, we were all trying to project and predict what our future would look like with all of these new space-aged communication gadgets, inventions, and methods claiming we will do even more work in less time.

At the same time, flashy degrees and office jobs were being sold as the gold standard. Construction was seen as a dropout career or a backup plan, a last resort. Media and culture began casting tradespeople not as hardworking, down-to-earth humans essential to the economy, but as dropouts—people who "couldn't cut it" elsewhere.

But that couldn't be further from the truth.

As someone who's worked in this field, earned advanced degrees, and spent time in both academic and executive circles, I can say this with confidence—and with a clear basis for comparison: construction folks have always been, and continue to be, some of the most competent, resilient, and sharpest people I know. Many of us may not have thrived in the classroom (myself included), but that doesn't mean we lack intelligence. Far from it.

Construction demands the ability to see both the big picture and the details. It takes critical thinking, spatial and social awareness, the ability to think and act under pressure, persistence, patience, and a whole bucket of logic and problem-solving. It also requires judgment, grit, and intuition—the kind you can't learn from a textbook. I'll continue to challenge anyone who says otherwise.

Looking back, it's clear that most of the "fun" in this era—like those before—stemmed from interaction and relationships: camaraderie, trust, integrity, and accountability.

There might also be some fun flakes buried in the rapid technological advancements, some of which show potential to solve some of our larger industry problems and challenges. But in this era it's still a bit too early to tell, and the general consensus about electronics and tech seems to be leaning closer to skepticism and uncertainty than fun.

A Hop, Skip, and Jump to 2000–2021 and Beyond

Let's fasten our tool belts and move into more familiar territory. If you've been in the industry for the past couple of decades, this won't be history—it'll feel more like muscle memory. From flip phones and Palm Pilots to pandemic pivots and software overload, this most recent stretch tested every part of who we are as builders.

While the rate of technological innovation in this era is staggering, let's focus on just one: the "smart" phone, introduced in 2007. In the early 2000s, you and your crew become more dependent on digital devices. So much so, in fact, that companies and clients pour hundreds of thousands of dollars into protecting systems from crashing when the clock strikes midnight on January 1, 2000 (of course, whose time zone that would be is anyone's guess). You're using light tables, digitizers, Nextels, and Palm Pilots. As a GC, you're renting out printed plan sets to subcontractors for takeoffs and estimates, which they fax back on bid day. Business cards are still a big deal—you've got binders full of them, organized by project with tabbed dividers for easy reference.

It's totally normal—and even good form—to bring a case of beer to the crew on Friday afternoons. You crack one open on the tailgate, check in, talk shop, and swap weekend plans. In the office, it's the same: folks gather around the break room or conference table—still warm from a heated change order debate hours earlier—to share a few beers and connect. If you're the new guy or gal, it's unofficial policy that you buy the first round with your first paycheck (handed to you in a sealed envelope that you have to take to the bank in person).

A few years go by, and you get promoted. Congrats! It comes with a heavier load of responsibility (thinly disguised as authority) and a tech bundle: a BlackBerry, laptop, and tablet, each with fifty different software programs for everything from QA/QC (Quality Assurance/Quality Control) to project management, none of which talk to each other. You also get a logoed polo shirt, a company truck, a fuel card, and a mug you'll never see again after your first site walk.

You're running meetings, clocking ten-hour days, and expected to always have your phone on—just in case of emergencies or 2 a.m. questions from someone who can't sleep. You're on site by 6:30 a.m., not having touched your computer and certainly not yet having made any progress on the steadily growing mountain of emails awaiting your response. Yet you're already getting follow-up calls from folks wondering why you haven't replied. It's clear you're increasingly expected to be available 24/7.

Construction is a team sport, and the game has just leveled up and become exceedingly more complex. Along with all the new gadgets and start-ups competing and trying to sell you on the "best new construction software" claiming to "fix all of construction's problems," the expectations, requirements, risks, and regulations are also proliferating. Everything demands your time—more meetings, more training, more trial and error—yet you're still expected to run your job and lead your team like nothing's changed.

Your teams consist of an array of skilled yet diverse individuals, ranging from those with PhDs to those who didn't graduate high school or who may have checkered pasts. You not only have to keep pace with the ever-changing requirements and certifications, you also have to be able to console and mentor the unskilled team members who aren't getting proper training and guidance on how to work on a team and do their work correctly. You have to be able to speak different languages (office, field, engineer, client, inspector, etc.) to resonate with the different personalities, backgrounds, and skill levels.

Cell phones are now basically body parts. By the early 2000s, personal digital assistants (PDAs) had pretty much replaced pagers—and then smartphones took over when the iPhone launched in 2007. Now, these devices hold everything: our calendars, contacts, passwords, reminders. We've offloaded a lot of our memory to them. And along the way, we've lost some of the face-to-face connection that used to happen more naturally. Jobsite conversations and water cooler updates have been replaced by emails, video calls, and conference lines. Even a regular phone call feels like a throwback.

All this to say, you've seen everyday work completely changed with a thousand small technological changes, *and* you're being asked to do more with your time. The added responsibility you feel builds and gains momentum like a gigantic ball rolling downhill. And there we are—at the bottom, arms and eyes wide open, smiles plastered on our faces for the sake of marketing, trying to convince ourselves that we are ready to catch it.

Spoiler: we're not.

Welcome to 2007–2012: The Great Recession.

It starts with the housing market collapse in 2007, triggering what we now call the Great Recession. Construction gets hit hard—projects stall, companies fold, and tradespeople across the country feel the ripple effects. The auto industry crashes, too, adding even more pressure. And while the downturn may have officially ended in 2009, the impact drags on for years—and in many ways, we're still carrying it today.

Not surprisingly, you (and I, and everyone else in the industry at the time) get crushed. But being the kind of people we are in this work, you get back up, dust yourself off, and start picking up the pieces. You're gritty, no doubt—but you're not invincible. You're doing all of this while limping, still healing from the impact yourself. And you're not just picking up what you were already carrying. You're also taking on the load left behind by teammates who were laid off—all in an effort to keep the company, the projects, and in many ways, society, from going under.

Some of your workmates aren't so lucky. They can't recover from a hit that big—and they leave the industry altogether, never to be seen or heard from again. Some of your subcontractors and owners on your current projects file for bankruptcy, causing gaping holes on projects and within teams, all of which must be infilled and carried out by the remaining team members. Lien waivers and joint checks become the new normal.

Those of you remaining seem to hold on for dear life for the next five to seven years without much relief from stress, expectations, or workloads. You notice construction executives and project leaders starting to shave their heads to account for the dramatic hair loss from this impactful era, creating a new acceptable fashion for those in our industry. For those who can't justify shaving their head, the added workload and anxiety about potentially losing their job creates unrelentingly tense muscles and hair follicles forcing every bit of hair out of their head anyway.

It's a difficult decade, but by around 2018 to 2019 you start to feel a bit of confidence return to your pace, stride, and workload. At last you are able to consolidate many of your devices into just a few multi-tasking tools, and your construction technology and software are starting to talk to each other and actually become more of a help than a hindrance. The economy is getting back on its feet.

But as soon as you have rebuilt momentum and are moving along at a faster yet finally manageable pace, March 2020 hits you in a head-on collision—with a global pandemic.

However, instead of the typical middle finger that you have come to expect from most of society throughout the past couple of decades, all of a sudden you get a standing ovation, smiles behind logoed face masks, and an imaginary, flimsy, plastic pin-on button stating you and your teammates are "essential."

Your neighbors get offered safe "work from home" solutions. While you—after talking to them from a distance about juggling kids, food shortages, homeschooling chaos, and the hoarding of toilet paper—head on out to work, where you keep water running, lights on, and buildings safe. Your commute is cut in half because the roads and highways are eerily dead—only "contractor white" pickup trucks with headache racks join you on the road. It all feels good for a while. You feel proud, important, and finally recognized for your impact and all your hard work.

None of this time was a joyride, and many of us were aboard the same struggle boat on rough seas. Either way, on top of it all, we were expected to keep pace with constantly changing government regulations and requirements that were tossed at us on a weekly basis.

As 2021 rolls around, the "We love our essential workers" signs are starting to fade and disintegrate into the soil. And just like that we're back to being the forgotten cogs that keep society running comfortably.

But, hey, our industry has made it through *yet another* economic nosedive. While each of these challenging eras has been tough, our industry as a whole has persevered. It had to. And along the way we gained valuable experiences and learned critical lessons that have strengthened our resilience and endurance. Our industry remains strong because of those of us who keep showing up and putting in the work.

This brings us, more or less, to today.

Before we wrap up our field trip from the 1890s to today, there's something important I want to leave you with. Over the past five to ten years, I've seen a shift. More people are beginning to understand who we are and appreciate what we do. Shop classes are returning to high schools. Career and Technical Education (CTE) is gaining traction. Youth apprenticeship programs and industry partnerships are helping students gain real-world experience before they even graduate. More young people are starting to see the trades as the legitimate and successful career path they truly are.

This shift hasn't happened by accident. It's the result of intentional effort from those of us in the industry to clarify, correct, and push back on outdated perceptions. And while that progress is encouraging, it's also still early. Many of us still don't feel fully appreciated for the amount of effort and work it takes to do what we do. The old stigma hasn't disappeared—not yet. But the momentum is real. Parents, educators, and communities are beginning to see the trades not as a fallback but as a worthy first choice, whether someone gets there through college or heads straight into the field.

After years of being looked down on or dismissed, there's finally a sense of recognition. A sense that maybe, just maybe, the work we do is starting to get the respect it's always deserved.

Still, even with that progress, we can't take our foot off the gas. It's going to take steady, ongoing effort from all of us. And the best way to do this? Keep showing up. Do the job with pride. Work with one another, not against each other. Remember who you are and all that you do every single day. Not everyone sees the long hours, the weather, the stress, or the weight of this job. But I do. And your crew sees it, too. That's what matters. Keep going.

Before we end this ride through history, I want to share one of my favorite quotes. It's from Theodore Roosevelt, from a speech he gave in Paris in 1910. It's about what it means to be in the arena. And I think it sums up exactly what we do every single day:

> It is not the critic who counts; not the man who points out how the strong man stumbles or where the doer of deeds could have done better. The credit belongs to the man who is actually in the arena, whose face is marred by dust and sweat and blood; who strives valiantly; who errs, and comes up short again and again, because there is no effort without error or shortcoming ... who at the best knows in the end the triumph of high achievement, and who at the worst, if he fails, at least he fails while daring greatly ...[18]

---

18 Theodore Roosevelt, "Citizenship in a Republic," transcript of speech delivered at the Sorbonne, Paris, April 23, 1910, https://www.presidency.ucsb.edu/documents/address-the-sorbonne-paris-france-citizenship-republic.

True to the cyclical nature of this industry, these past couple of decades have been gnarly in their own way. Digital technology is now considered essential to our work and is constantly reshaping how we manage our teams, track progress, and communicate. All the while, societal and regulatory expectations have stacked up to an unmanageable height. At the same time, efficiency has become the new gold standard, pulling much of our focus and energy.

Our focus is knocked a bit out of level with all the expectations and requirements thrown at us. But this era, like all of those preceding this one, has challenges that the industry has and will overcome. How quickly we overcome them—and if we come out stronger or weaker on the other side—is yet to be determined and lies in the hands of you, me, and all our fellow industry members.

We didn't always know what was coming, but we kept showing up. Kept figuring it out. And while it hasn't always felt fun, there's something powerful in doing work that matters—especially when you are doing it with people you trust and respect.

Our Bus Ride Back to Reality

Our field trip through construction's past is over. Let's take a look at what we've collected.

1890s construction work wasn't so different from our current work environment as many of us would have thought! Yes, technology and equipment have most definitely advanced since then, but at its core—the interactions and people aspect of our work—our industry today isn't *that* different.

Wyckoff expressed that he felt connected, welcomed, and accepted as a part of the pack during this work, which is an essential component to enjoying what we do. We will unpack this further later in the book, but this may be his hint to us that these are the pieces of the fun puzzle we are looking for:

> I have received the strangest kindness from the men, most tactful treatment of me as a novice. They laughed at my strenuous efforts to do what was so much easier to them, and they laughed when the boss singled me out for abuse, but never ill-naturedly, I thought. And those who made up to me, and with whom I picked up acquaintance, showed the kindest consideration... In spite of their roughness and hardness of manner and speech, one never felt the smallest fear of these men, and you had growing feeling that their better natures were never far to seek.[19]

Another potential lesson we can glean from Wyckoff's experience is to watch our "my-way-or-the-highway" tendencies that may suck the fun out of an experience. While physical labor had the potential for fulfillment, he expressed that this potential was quickly extinguished by how the work was being micromanaged and dictated by the boss. So, while the physical aspect of nineteenth-century construction was relentless, working alongside his workmates, on the other hand, would have been a hoot! Between the lines, the fun is still there—in the camaraderie, the shared respect, and the fulfillment from a job done well.

The 1920s and 1930s saw a big swing from boom to bust which affected everyone. This era holds lessons about the importance of a growth mindset and of personal and professional development: if we allow ourselves to become too comfortable or overly cocky, then we are at risk of becoming stagnant or chewed up and spit out by the industry. We need

---

19 Wyckoff, *The Workers*, 73.

to intentionally invest in continually improving ourselves, to ensure we have the strength or endurance required to get into the ring to face the potentially devastating punches the economy inevitably throws our way.

Pride became an important aspect of our work, too, thanks in no small part to publications sharing images of the Sky Boys. Even with the unrelenting stress of keeping their jobs and supporting their families as the economy crumbled and fell apart around them, their faces reflect beaming pride—a key component in fulfillment. It is not only a sense of pride in the work you are doing, but also pride in who you are and what you are a part of and contributing to. Again, the work was difficult, but fulfillment could still be found.

The middle of the twentieth century saw a huge rate of growth for our industry, fed by production needs of the World War II war effort and the post-War boom. Construction ramped up with newly minted power tools, heavy equipment, and a surge of suburban growth. Women entered the workforce at record rates, shifting jobsite dynamics and proving they were more than ready to tackle any role. Meanwhile, gadgets like Tupperware, push-button phones, and pocket calculators offered small glimpses of modern conveniences and efficiencies, even as OSHA and environmental regulations added new layers of responsibility. Yet even with all these changes—more people, bigger equipment, and tighter rules—many still found pride, camaraderie, and even fun in their work.

In the 1980s and 1990s, we saw more digital technology make its way into construction, but the industry remained reliant on the humans, just as it always had. New hurdles appeared here, too, with construction work becoming perceived as a "dropout" career path and GCs getting scrappy over dollars, risk, and ever-shortening deadlines. While the day-to-day challenges hadn't changed, how we approached our work became more difficult. This seems to have followed us into the early 2000s, with the pressure of the Great Recession forcing more responsibilities onto fewer people, building tension. Camaraderie seemed harder to build and maintain, as there was more to do and less time to do it in, and cell phones and computers were increasingly pushing out in-person interactions. Then as we headed toward the 2020s, new challenges threatened how and why we work. Still, there isn't a clear point where "fun" disappears.

While many components of these eras may have seemed slightly less than enjoyable, there are many hints and answers speckled throughout them, guiding us on how we can inject fun back into our workdays and careers.

I'd like you to stop for a moment.

- Close your eyes.
- Take a large mental step back.
- Take a few deep breaths.
- Do a quick mental fly-through of the construction industry's timeline that we just covered.
- Open your eyes again.

Consider all that construction has been challenged with yet persevered and adapted to over the past century and a half. The challenges we have overcome are not only incredibly impressive but also encouraging! Recognizing and appreciating this is important.

While, yes, these challenges can be a massive pain in the butt, they can also be blessings in disguise. It's the challenges that attract and retain the type of people equipped and willing to tackle and endure these economic plunges, feats of engineering, visionary designs, skewed societal perceptions, and unrelenting project challenges.

If you grab a mental magnifying glass and take a closer look at what the construction industry is and what it is made of, it's an amalgamation of all the people who have previously worked and are currently working within it. It's you, it's me, it's Walter Wyckoff from 1890 and the Sky Boys from the 1930s, and yes—the tech-savvy visionaries of the 2000s and beyond. It's the mindsets, actions, decisions, energy, and effort of this massive conglomerate of individuals. It's all of us who contribute to the larger overall experience and reputation of our industry for those within and outside of it. We are gritty, intelligent, loyal, strong, adaptable, and resilient. We are construction.

Construction has and will continuously encounter challenges. This means we, as individuals and teams, will continuously encounter challenges, yet because of the type of people we are, we will continue to fight! We will absolutely get knocked on our butts every so often, but we are the ones who get back up, dust ourselves off, and, with a bit more experience and determination, continue fighting for what we know is essential and important to our communities and the world. I hope you are as proud as I am to be a part of this industry and work alongside these incredible people.

The inherently fast-paced, high-pressure nature of our industry frequently makes it easy to forget and miss these valuable fun nuggets when we are speed-walking past them or are buried and lost in the weeds of our day-to-day tasks. However, if we stop every so often and step closer to find these gems or even step back to look at the larger picture, they become easier to notice and pick up.

While I cannot create or define "fun" for you, the rest of this book is dedicated to providing you with some navigational skills and tools, as well as guidance along the proverbial industry map to show where fun is more likely to be excavated and rediscovered—and how to harvest and ignite it.

Our historical field trip has provided a helpful background and a wider lens to better identify what the deeper root cause may be to our fun deficiency in the industry. It also provides us with a better understanding of where we come from and who we are as an industry today. Before we can provide proposed solutions to this industry-wide problem, we have to excavate a bit further to see if there is any one main contributing factor to eliminating our access to fun.

In the next chapter, we'll dig through our collected findings and glean insights and perspectives from some of our current industry colleagues to see whether we can identify any consistencies or glaring culprits.

GUEST SPEAKER, AUGUST 2024

"Do you think construction is fun?" I ask as I lean closer to my computer screen, which projects a live video of a seasoned industry colleague sitting almost 2000 miles from me.

John*[20] is a respected member of the construction industry who entered the industry in 1977, five decades ago. Today, he's wearing a loose, comfortable, grey-blue three-button polo instead of his more typical executive attire of a crisply ironed shirt and fitted sports jacket. The office scene behind him is equally comfortable and familiar. A glass markerboard with scribbled notes, a particleboard bookcase placed alongside it holding industry publications and reference manuals, and a canvas print displaying an action shot of two three-axle articulated CAT haul trucks being loaded by an excavator somewhere in the mountains complete the décor.

"I do. Camaraderie, achievement, you know? We're doing things that have to be done for society. I think there is just tremendous purpose in it. That's one of the reasons I think it is fun. If you get a true sense of purpose from something, you can go through a lot of bad days and still call it fun."

"So, looking at it from a different perspective, why do you suppose others within our industry are saying it isn't fun anymore? Why would they be saying this?"

He pauses, crosses his arms, and starts rubbing his cleanly shaven chin, allowing himself the slightest amount of time to decide on the best words and tone to properly convey his message. It's an admirable technique born from years and

---

20 Some names have been changed to respect individuals' privacy. When used, an asterisk (*) indicates a pseudonym substituted for the person's real name.

years' experience of difficult discussions and having to communicate in the many languages of construction.

Uncrossing his arms, he explains: "There are enough hurdles to enjoyment, you know? The weather can be terrible, the traffic can be terrible—even though you are trying to fix a road so that people can have a more efficient road, while you're doing it they're cursing at you as they drive by. There's a lot of that, and then the conditions aren't always what are laid out, so you've got those challenges. And then the inspectors aren't easy to get along with. They want to fight with you instead of moving the job ahead. People don't show up when they're supposed to; they don't do things when they say they're going to. It's all these things, right?"

John's "fun hurdle" examples kick up strong memories of similar experiences, and I find myself laughing and nodding along.

"If everybody would just live by, roughly, the same set of values... Like, if I tell you I am going to do something, I am going to do it. And if everybody lived by that value, then I think it would bring the fun back!"

I ask him whether he thinks there was a point in time when this was the case, when people were living by some of these same values, and when others may have considered the industry fun.

"I honestly think there was." He pauses to adjust his glasses. "I can tell you this: in the late '80s to early '90s, the general contracting world got really, really dirty. It got *really* dirty." He pauses and shifts slightly in his seat. "Bid shopping, abuse of subcontractors; any way they could get a project because they were working so hard to get work."

I'm surprised. "Why do you think it was in the '80s and '90s that this all started or when this was going on?" I ask. "What was happening?"

"The construction economy was abysmal at that time. A really bad economy—'90, '91, '92. Not enough work, and they had to get their margin where they could get their margin, and they took it out of the subs. In 2000 and in the Great Recession, the construction economy got to be a bit nasty as well, but the late '80s and early '90s were much worse."

My interview with John has told me many things—but whether it was ever "fun" isn't yet clear. We'll have to continue searching.

3

Locating the Proverbial Leak

"Out of intense complexities,
intense simplicities emerge."
WINSTON CHURCHILL

IF YOU'VE EVER picked up a new smartphone, had a knee replacement, or earned any type of industry certification, you know that even when you get a new gadget or have a few parts replaced and improved over the years, it doesn't change the core of who you are. In the same way, industry gadgets, tools, and techniques come and go, but they don't fundamentally change what construction is.

Our jaunt through history has shown us that the heart and core of who and what construction is really hasn't changed that much, even though many of the gadgets and tools have. If anything, the advancements in technology and techniques over the decades—by far the biggest changes—are just infilling sections of our festering labor shortages that have built up for far too many years and have caused an inconvenient limp. At its foundation, this industry still runs on grit, skill, and teamwork—just like it did when Walter Wyckoff gave our industry a go back in the 1890s.

To the untrained eye, our field trip along construction's past can look a bit rough and unpromising. On the surface, there aren't any glaring answers (leaks) to our "fun" problem (soggy drywall). But, then again, in our industry we aren't known to take problems at surface value and give up easily. We need to take a closer look and dig in a bit further to find the root cause of our leak.

The last chapter was built from research—flipping through antique pages of industry magazines, journals, and books. But to uncover what's been going on over the past few decades, I knew I had to go straight to the source: our people who have or have had boots on the ground. The people who know this work inside and out. To best identify where the fun went, and how to bring it back, we need to ask and listen to those who've lived it.

Insight from Those Who've Walked the Walk

As part of my research for this book, I sat down and talked with a range of respected professionals across the industry. Some had a decade of experience under their belts. Others had been building, leading, and problem-solving in construction for over forty-five years.

As we talked, one common theme kept coming up with nearly all of them, no matter their role, background, length of industry experience, or opinion: they had heard the comment,

"This just isn't fun anymore," echoing through job trailers, break rooms, and boardrooms.

So I followed up by asking them the same simple but telling questions:

1. Do you think construction is fun? Why, or why not?
2. If it used to be fun, when was that? And when did it stop being fun?
3. What made (or makes) it fun, and what would make construction fun again?

I dug into their responses, looking for patterns, common threads, and hard truths. What follows isn't just data. It's the honest perspective of real people who live and breathe this work every day.

Let's hear what they had to say.

Do You Think Construction Is Fun? Why or Why Not?

Ask around and most folks in our industry will tell you the same thing: construction can be tough, no doubt. But when you're part of a solid team, building something real, and staying connected to why you got into it in the first place, it's also one of the most rewarding (and, yes, *fun*) careers out there. A lot of that enjoyment comes from both what we bring to the work—our mindset and our pride—and the people we work alongside.

A sense of freedom and of challenge came up more than once. One superintendent with a large GC, with twelve years of experience, put it well: "Who is lucky enough to have freedom in leadership? I don't like assembly-line work. I like that the freedom and path is different in everything I do. You get new co-workers every project, so you get a break from any crappy ones. There is also no shortage of problems to solve."

Unsurprisingly, camaraderie and strong teams came up again and again as the key to making tough days more enjoyable. "The most difficult projects are my favorite," one industry professional shared, "because we come together as a team. We work with amazing and incredibly smart people from all walks of life, and it can be fun along the way. It's fun to be a part of something rewarding."

The sense of belonging and contributing to something bigger was a common thread. The president of an electrical company with twenty-one years in the industry put it simply: "Feeling like part of a team; part of the system is fun." And a former electrical project manager with thirty-eight years in the field echoed that, saying, "Seeing something created out of nothing is fun, and the camaraderie of the workers, even after challenging times, can make it fun."

Others linked the idea of fun more to a sense of personal responsibility and to a mindset. "There were always fun aspects throughout my career," said a principal architect with twenty-nine years of experience. "If you don't have relationships, it isn't fun. But also, we think it isn't fun because of something outside ourselves in lieu of something inside ourselves."

"I'm a big believer in making things fun," said a workforce development manager (and former superintendent and project manager) with twenty-six years in the industry. "I have to go out of my way to make it fun. When I was in a superintendent role, I had to earmark time to make it fun. It's our responsibility to make it fun. It is the least fun when I was furthest away from why I got into this industry."

"I had more fun in the field," said one heavy civil CEO with forty-eight years under his belt. "When you get into management there is more bureaucracy. But it is still fun. The camaraderie, achievement, and purpose. When you have a true sense of purpose you can make it through a lot of bad days. Fun is different for everyone. It is our responsibility to make it fun."

But not everyone resonated with the word "fun" the same way. The vice-president and division manager of a national GC with two decades in the industry offered this: "It is a very rewarding business. There are very few industries where you can see your results and feel ownership and pride in how it got there. There is a misconception that construction is hard. The truth is that success is hard. It takes work and dedication. Construction isn't a 'fall back' career anymore. It's a big business, not something we do on the side. This is my career, and my expectation is that I am here to work. It's not a birthday party."

A Marine Corps veteran and now-CEO with fourteen years in construction simply said, "I don't think that 'fun' is the narrative of the industry right now." And the general manager of a mechanical contractor with forty years of experience didn't warm to the idea of fun at all. "I don't know if I would call it fun," he said. "'Fun' sounds trite."

In summary, many enjoy the variety, changing projects, and problem-solving nature of the work. Fun is strongly linked to the strong teams and relationships on jobsites, allowing teams to work together and achieve results.

If Construction Used to Be Fun, When Was That? And When Did It Stop Being Fun?

Another recurring theme emerged: the decline in teamwork, trust, and shared purpose has contributed to making the work feel less rewarding in recent years.

A general superintendent with thirty-one years of experience said, "It was fun when everyone worked together! They just jumped in and did it without being asked. The goal was the project, and we all gained financially by working together. Money should be the byproduct, but relationships should be the reason."

While many had said earlier that the people were the best part of the job, some pointed out that camaraderie and connection have eroded. One shared that "managing personalities and egos" is a challenge. "As soon as people want to make it their own and think they are above everyone else, it's a problem. It's [not] fun because they think they can change other people. It isn't fun when we have to argue with everyone all the time."

"It used to be fun," said an individual who'd seen a great deal over three decades. "Construction used to provide a brotherhood mentality, but now there is a lack of camaraderie. Or a lack of opportunity or time to build camaraderie." He went on. "We used to have time to talk through things and build camaraderie and work out problems and challenges, but this is gone. We hit the ground running, and are expected to keep home at home and work at work. No time for coffee, no time for beer, and no time to just talk and work through things. Work requires a lot of focus, so you can't focus on home stuff, but there is no time to work out home stuff anymore either."

This was echoed by another seasoned professional. "One, the general lack of trust and respect of our knowledge, skill, and experience by those outside of our immediate industry," he said. "Two, technology plays a part in our industry not being fun anymore. So many people are keyboard bullies instead of having to work it out in person and build workplace relationships. Three, the speed of construction at all costs. Speed takes a toll on employees both in the office and in the field, and then takes a toll on the family units. And for what? It's hard to get people out after work to hang out or socialize. They all rush home to family. No time, lack of time for some reason."

Several folks could pinpoint the moment when the fun started fading. "It was fun up until 2008, when we lost all the people that knew what they were doing," said one superintendent. "We had to fill roles (because of the economy taking another dump), so we just filled them, but not for ethical purposes. We stopped training and mentoring at that point. It was our job to do that, and we didn't. We weren't taught how to. It was a lot easier to say, "It's not my problem.'"

As mentioned in the breakout box in Chapter 2, one industry veteran spoke about shifts starting in the late 1980s and early 1990s, particularly in how GCs operated. In our interview, he expanded on how communication had changed in the digital age. "Before 2000, there was a lot more intentionality around communication. You had to plan your conversations, because you had to be in your office on a landline. Communication has become hollow, because of the convenience of it."

Another added that it was fun "when we felt a sense of accomplishment together. When it was less bureaucratic, and we could just go out and do the work. Now we have a lot more data, rules, and restrictions."

"Society has become so litigious," said one, who went on to explain how she felt she spent more time "covering my ass on paper instead of working it out in person through relationships. What is our motivation? To stay out of court, or to build and design amazing things? I don't want to have to document every word."

She continued, "This is taking us away from what we originally got into this industry to do. My dad was an architect when drawing sheet sets were twenty-five sheets. I was brought to meetings, and he would work with the contractors in-person and build relationships. Now our focus is so siloed. We say the world has changed and because of that I am unhappy. But the world will always change. I have efficacy over whether I'm happy or not. You have to understand yourself to know what gives you energy versus what takes it."

Some felt that camaraderie had eroded to the point where healthy conflict had become impossible. "People aren't taking time to dig in and discover anymore," said one executive director. "They just want to complain, and if they don't get what they want, instead of working it out with their supervisor, they keep going higher until they get what they want. Confronting issues seems to be a thing of the past. Instead, they avoid or prolong the conflict."

Alongside concerns about culture and communication, many also pointed to increasing bureaucracy as a major source of frustration. Paperwork, red tape, and what one person called "the number of unproductive meetings" were wearing people down. Others mentioned how "large general contractors are politicizing the jobsite," and how "overgeneralized DEI [diversity, equity, and inclusion] mandates, and a focus on pointing out differences, cause more division."

These additional responsibilities weren't just time-consuming, they were creating more friction in an already demanding job. As one person put it, "There's too much finger-wagging instead of ownership and accountability."

In addition, one interviewee, an executive director of facilities management and owners' rep, pointed out how cost pressures have added a new layer of stress. "There's no more tolerance for mistakes, because we're too afraid of added cost. Not all mistakes lead to cost. We can't only focus on finances and cost. We need more RFIs [requests for information] and fewer ASIs [architect's supplemental instructions]." His comments also reinforced the importance of shifting toward a bottom-up approach—encouraging more collaboration, communication, and decision-making from those closest to the work, rather than relying solely on top-down directives.

Not everyone sees things the same way. One interviewee shared, "A lot of people are leaving and retiring. When they leave, it isn't fun. We are in the middle trying to navigate it all. We now have to step up and be leaders, but not everyone is cut out to be leaders."

Others see signs of progress instead. "It wasn't fun, but the past ten years have been more fun. New, fresh, and innovative ideas are coming to the table and creating a new way of doing business in construction that is more about creating relationships and transformative spaces—from civil to vertical construction."

One summed it up this way: "It's no more or less fun. It's just different. Change is difficult for everyone. Nothing will ever be as good as 'the good ol' days.' Construction is a pay-me-now-or-pay-me-later industry. If you're not working hard and doing what you are supposed to be doing now, you will pay later, and it won't be fun."

And finally: "We all play an integral part in the larger machine. If most of your attention is focused on the bad in our industry, you're missing all the good. There is a lot of good in our industry."

In summary, many of those interviewed felt that the industry used to carry a stronger sense of camaraderie and brotherhood, elements that made room for fun. Even on hard days. Today, the fast pace, added bureaucracy, and shifting expectations can make that connection harder to find. And while technology has made much of the work easier, it's also made our relationships more transactional and less personal.

What Made (or Makes) Construction Fun, and What Would Make It Fun Again?

For the final question, I asked what we could actually *do* to reignite or repair the sense of fun back into the industry, or to keep it alive where it still exists.

Construction workers are part of a long and respected history, and our work, when done well, can last for generations. "Appreciating and understanding what it means to 'work as a team,'" said a GC superintendent. "It is a trade, and how honorable it is to have a skill and be a part of a trade."

Several responses pointed to a team's shared purpose as a key driver of fun and fulfillment. "Bring team members into the plan," said a company president. "It's their baby, too. A 'we', not 'me' mentality. Things that are challenging are fun, but we have to have some freedom. Yes, we all need goalposts and rules, but allow people to contribute within those rules and be involved in something exciting. This allows people to have pride in their work."

Others emphasized respect and recognition as essential parts of the equation. "There's a perception that we should build it faster with the same amount of work," said a GM, reflecting on what wasn't working. "Also being treated as a 'less than.' Still having to use unsanitary Port-O-Lets when the restrooms have been finished on a project. Or having to work 7-12s [i.e., twelve hours a day, seven days a week] with no regard for the workers, just because the owner wants it done." This weakens the opportunity to take pride in your work and have fun while doing it. "Respect for industry experience" is important, said another. "The 'everything is the same and everyone's the same' mentality needs to go away."

An industry executive emphasized the importance of human connection, reminding us that we're not just managing tasks, we're working with people. "Personalization of communication and commitments. Reintroducing care for one another and humanizing and personalizing what we do. Do the right thing, no matter what!"

This view was reinforced by a superintendent: "One: fairness. Treat others how you want to be treated. Or even more simply, treat people as people. Two: ownership and involvement. You change the industry by changing yourself. Construction people are some of the best people."

"Have conversations," said another interviewee. "There are a lot of emotions in construction. If you can learn to understand conversations, you start to understand and appreciate the emotions behind them."

For many, the sense of fun is directly tied to seeing results and watching ideas come to life. "Fun is the sense of accomplishment and celebration when you hit milestones. The tangible aspect of your hard work. When it is finally visible, it's that feeling of, 'Holy shit! This is cool!'" relayed one interviewee.

A workforce built on trust and collaboration rather than blame makes the work not only more fun, but more sustainable. "Allow more time for work to be put in place, and this will mean fewer mistakes. Reduce blame and increase collaboration, planning, and respect."

Ultimately, the people I interviewed understood that people have to come first. "Connect with others and help them feel like a valued team member. Share a passion of a vision made into reality. It can't be transactional; you need to be recognized as a partner. Understand that we will have tough times, but being transparent will help us through them and be engaged as a team, where you get to share ideas, experiences, and learn collectively. You can be proud of any accomplishment."

Mindset, above everything else, seems to determine how we experience our work in construction. "Change our mindsets and perspectives," said an executive director. A VP agreed, adding, "A better understanding of expectations and defining what 'fun' is for you." He continued: "Personal ownership is important: if it's not fun, then make it fun. If you are relying on others to make it fun, you will wait a long time. The people who are making it fun for themselves typically make it fun for others and wind up being way more successful."

At the end of the day, it all comes down to purpose, pride, ownership, and respect. When your team feels like they're working toward something meaningful, and their hard work is noticed and appreciated, that's when the job starts to feel fun again. Being included in the plan, having a say in how things get done—when your knowledge, experience, and perspective matter—and knowing you're trusted to do quality work, that's what makes the difference.

It's about more than swinging hammers or running wire. It's about building something that lasts, side-by-side with people you trust. The ones who take ownership, who find ways to enjoy the grind, are often the ones who lift everyone around them. Fun doesn't just show up. You create it by showing up, working hard, backing each other up, and taking pride in what you build together.

The Leak Causing Our Soggy Fun Problem

Our construction history field trip and talking shop with fellow industry members helped us excavate the essential element behind fulfillment and enjoyment in our industry: the human element. More specifically, it's the connections, welds, and adhesives of the pipelines that link us together. And lately, those pipelines have been stretched, overloaded, and eroded, causing leaks.

The reality is that there probably isn't one exact moment we can all point to and say, "That's when construction stopped being fun." Sure, there were rough eras that strained resources, overloaded teams, or made it harder to connect. But as we've heard from those within the industry—and maybe even recognized ourselves—fun doesn't just happen. It's not a feature of the jobsite or the office. It's something we create, with and through our people.

The faster pace of projects, mounting expectations from clients and upper management, and the siloed specialization of our roles have likely caused punctures and weathering in these critical human pipelines. The quality of these connections has deteriorated and, with it, our sense of fun. And somewhere along the way efficiency became so shiny and glamorous that we stopped noticing the infrastructure it was quietly eroding.

All of this has resulted in quite a few fun leaks, causing our pipelines to look more like perforated pipe than solid pipe, thereby causing the underlying leak to our soggy fun problem. One drip, one day, one overlooked connection at a time.

For EFFiciency's Sake

Let's be clear: efficiency isn't the enemy, and you don't have to choose between being efficient and being human. Heck, efficiency is welded to my core—it's a value stored down in my gut whether I like it or not. Turning it off, for me, isn't an option.

My brain automatically starts trying to figure out how to do a job faster, better, or with fewer headaches. It's just how I'm wired. And yeah, I get irritated when I see a clunky process that no one's fixing. So trust me, I get the pull of efficiency! But that's also why I'm saying this: maybe we've zoomed in on it a little too much.

In construction we've locked in so tightly on speed and productivity that other crucial elements have faded into the background. And it's no surprise because efficiency feels good. It's measurable, controllable, and more tangible and concrete.

The human element? Not so much. It can't be boiled down to a formula or a single line item. It's more challenging, unpredictable, and a whole lot easier to push aside. But here's the thing: people are the most valuable, rewarding, and complex part of this business. And when we neglect them? No RFI, ASI, or lean tool can fix what breaks. We can't control other people; we can only influence them, and even this is never a sure thing.

I've seen and experienced what happens when the human side gets left out. You may hit your deadlines, but if your team's running on fumes—or worse, running for the door—then it's not really a win. So here's the deal: a healthy jobsite, a strong company, a thriving crew all come down to balance. You can't lean fully into efficiency and ignore your people. Doing that won't work for long. Sometimes you've got to take a step back and make sure the scales aren't tipping too far in one direction.

Now we know where the leak is. We've cut through the drywall, traced the source, and found the damage. Our human pipelines are eroding, and too often in the name of efficiency.

What would any respectable contractor or builder do next? We fix it! We make a plan, gather the right tools, and get to work.

JAKE'S JOURNEY

I have had the privilege of meeting some truly exceptional humans in our industry, one of which is someone who I very much respect. Over the past thirty years, he's faced and overcome not only the external pressures of our industry and economy, but also some incredibly tough personal obstacles.

If I had to guess who'd be most likely to call construction "not fun anymore," or even leave the industry altogether, I might've said it would be him. But he hasn't left. He's stayed. And somewhere along the way, he's rediscovered the fun in it. The mindset shift he made and the challenges he overcame offer up a kind of wisdom I think a lot of us can learn from—especially those who are struggling right now and wondering if they should stick it out.

Jake* is a senior superintendent for a large GC with over thirty years of experience. He and I connected a couple of years ago, and have since become good industry friends. His perspective and viewpoint of the industry are promising and inspiring, and I greatly admire his grit and tenacity.

I've never told him this, but he's become a bit of an industry mentor and role model to me. If I ever did tell him, he'd probably shift in his seat, scoff, and mutter that I need to get my head checked. That's just who he is. What he doesn't see—what he's too humble to admit—is how many people he's impacted over the years. He's one of those people who quietly changes jobsite culture for the better just by being there. But he didn't always show up that way. In fact, for much of his career, it was the opposite—until one afternoon in 2018.

Jake's comfort zone is definitely on the jobsite. He would've preferred us to meet in his trailer, surrounded by hard hats and steel-toe boots. But I needed to actually hear him—without the usual interruptions. I appreciated that he humored me by meeting at a coffee shop, even though he rarely drinks coffee.

We sat at an outdoor table, the only ones willing (or stubborn enough) to brave the chilly weather. He wore his usual short-sleeved plaid button-up shirt, jeans, and boots caked in dried concrete. Classic Jake.

When we started talking about his story, he paused, looked away for a moment, and said: "This is really hard, when I start talking about this."

That one sentence set the tone for everything that came next.

Entering the Industry

Jake took a breath, then started sharing. Right out of high school he enrolled in college, planning to major in elementary education. He'd been inspired by one of his own elementary school teachers, someone who encouraged him to become a strong male role model for kids growing up without father figures. That became Jake's dream, to make a difference, to give back, and help shape the next generation.

Around the same time, his first son was born. As a young adult suddenly responsible for supporting a wife and a newborn, Jake took a job in construction while still trying to keep up with classes. This was 1994, and he was just eighteen years old.

He kept that pace up for about a year. But eventually he had to make one of those tough life choices to put the dream on hold to take care of his family the way he needed and wanted to. He left college and went into construction full-time. "I lost the desire to be a teacher and role model," he told me, "and had to focus more on paying the bills."

Jake was very talented and motivated, and worked quickly into a carpenter role. He became a master of all trades, capable of doing everything from pouring and placing concrete to finish carpentry.

"It wasn't as specialized as it is now, where you have drywallers, concrete, finishers, trim-carpenters, rough-carpenters. I was able to do everything all the way through. I had some of the best teachers of my life during that time."

Hyper-focused on finding success and moving up through the ranks, his focus was primarily on work and clocking as many hours as possible.

He paused to consider before continuing. “I got into addiction the first ten years of my career in the industry, thinking that my addiction made it so that I could work longer hours to provide more, do more, learn more... I wasn’t the best.”

In 2008, he successfully earned the role of superintendent. This was bittersweet, as although he was achieving immense success at work, the pace, energy, and drug addiction were all taking an immense toll on him. “This is when I went through a divorce and went through rehab. During detox, my body shut down. I woke up in a hospital two days later.”

Starting Again

After rehab, Jake decided to start over. He hired on with a different construction company as a laborer—by choice. He wanted to start fresh and prove to himself that he could work up through the ranks. This coincided with the Great Recession, when the economy was unforgiving. Jake saw it as a formidable challenge, but also an opportunity to “step it up and be a little bit better.”

Unsurprisingly, Jake worked back up through the ranks very quickly and became a superintendent once again around 2010, this time without the crutch of drugs or alcohol. This quick advancement was one hundred percent him: his brain, his resilience, his fortitude, his skill, and his perseverance.

Once in the role of superintendent again, he found he was in need of his old mentors for advice and guidance on undertaking this new leadership and mentoring role. He quickly realized that all of them had retired or left the industry during the recession. It was now his role to teach other people, but he didn’t know how to do this and didn’t have the role models to coach him on how to navigate these new challenges.

Without guidance or coaching on how to lead, like many in our industry, he simply defaulted to a command-style leadership.

"All I did is complain about all these people around here that don't know what's going on. There are no carpenters, there are no engineers, there are no superintendents. I just did my own thing." Jake admitted, "The way that I looked at it is that, if you don't want to be a part of it, if you don't know how to do it, then just get out of my way and I'll get it done, I don't need you here. This was exactly the opposite attitude of what I went into my professional career for: to be a role model to people."

I, myself, took on this attitude around this time and noticed many other colleagues in the industry doing the same. During our trip through construction's past, we mentioned the fact that many of us who remained in the industry during the Great Recession were handed all the projects, responsibilities, and tasks previously assigned to counterparts who had been laid off. This was without allocating more time to do it, and for many of us it also came with a pay cut to ensure the company could stay in business. These added pressures made it exceedingly hard to engage in productive leadership styles rather than the survival leadership styles that aren't known to be conducive to great problem-solving or fostering the patience to teach others.

"I did this for about 10 years," Jake continued. "Just being spiteful in the industry. Not really caring about anyone around me and what they were actually dealing with in their lives. Not exposing anybody to what I had going on in my life. I got really high up in a couple of companies, but just didn't really care for anything at this time. I went from wanting to be a role model to being the worst role model I could be. Not caring about anyone around me, including my family. Just doing everything for me, trying to get ahead. And I did get ahead, to some extent."

This command-style leadership can be tricky to navigate in our industry, though. While it's effective in certain situations, especially short-term, as Jake quickly realized, it can do more harm than good if it's overused or applied in the wrong context.

On paper, Jake was one of the best superintendents around. He was bringing projects in months ahead of schedule and under budget. This is indeed lucrative for the financial and marketing aspects of a company. But it was at a large cost to him personally. While he was meeting all of these milestones and achieving professional success on the outside, on the inside he was unsettled and deeply unhappy. Without guidance or coaching from his mentors and teachers, his only feedback was the reward of promotions and being assigned large, complicated projects. He became numb and kept his command-style leadership, thinking this was the only way to keep advancing and become successful in construction. That is, until that one afternoon in 2018.

Revelation on the Scaffold

One afternoon in 2018, Jake found himself sitting on a scaffold next to a young man on his project who confided that he didn't want to or intend to live past the end of the week. Jake turned off his phone and sat on that scaffold talking to him for over two hours.

"I still talk to that young man today!" Jake revealed. "I had a light go on that day. To truly make a career fun or exciting, it's more than just the industry—it's the human element."

He explained further: "We sit here and complain about the human element, complain about the people. But if you don't actually care about the people you're working with, then, yeah, it makes it a pretty miserable industry. Everyone fails every day, everyone makes a mistake, every job has a problem. If you can't figure out how to fix that problem or work through that problem, then that's an issue. But if you can't even attempt to work through that problem or fix that problem, and you just put it off on someone else, then who is the problem? Every problem is fixable, if you want to fix it. You determine the cost. Is the cost the human element, or is the cost the simple deal of the dollar?"

As Jake puts it, the human element is key to fun, fulfillment, and enjoyment in our industry. It is what he searched for and chased after all those years; the connections and relationships of the people you work with and alongside provide meaning.

Instead of the 'look at me, look what I did' attitude, I now care more about people knowing my failures than I do about people knowing my successes. I don't ever want to be a 'look at me' person. I'd rather be a 'look at us, look at what the team did' person.

"I have been fired, I have been told I was the worst person in the world, I've been divorced, I've been a drug addict, never finished college, not 'educated.' Is that successful? I don't know. For me, I would consider myself successful; I am still standing. I now take my time, any opportunity I have, to to teach people, and to sit and listen to people. You don't know their situations when they arrive. It may mean that I call my current and forever wife and say I'm staying late, and this is why. I'll take calls from guys at 2:30 in the morning and meet wherever they need to meet."

Jake took a deep breath. It is hard for me to communicate my immense respect and admiration for Jake, my gratitude that he shared his story with me, with us—for the sake of helping and improving all of us in our search for fun and fulfillment in this industry.

My coffee was long gone by this time, and the outdoor temperatures hadn't budged sitting outside of that coffee shop. Yet I was not cold. I felt grateful, energized, and hopeful from Jake's outpouring of his life lessons and story.

His journey hadn't been an easy one, and I know a lot of industry members have similar stories or are in similar trenches of these life and industry challenges. But Jake's story provides inspiring and encouraging lessons about how we have control over our destiny. It shows that fun and fulfillment can be found by anyone who is looking for it. It isn't always an easy path, but if Jake can turn his mindset and perspective around and find fulfillment and fun in this industry, then any and all of us can. His findings along his journey have led us to a substantial discovery about what can make our industry fun and fulfilling again.

To make it fun, you have to dig down to the human element.

Jake uncovered this larger construction problem and solution the hard way, and largely by himself. Since this discovery, he has shifted his mindset to successfully find fun and fulfillment in our industry again.

Truth be told, I feel like a tantrum-throwing toddler for complaining about some of my challenges in construction. While I have struggled with my fair share of issues, including deep depression, stress-induced hospital visits, and the debilitating loss of fellow workmates, Jake's story and path reminded me that others carry heavy loads, too, and even the most substantial challenges can be navigated and overcome. This is one of those times when I could either continue down the "woe is me" path or the "suck it up and do something about it" path—not only for myself, but for all those I interact and work with.

This book is proof of the path I have chosen. I am forever grateful to Jake for sharing his story with me and allowing me to share it with all of you. I hope now you can see why I respect and admire him so much. He has discovered, the hard way, what many of us struggle with and are searching for, but he has also persevered and has found fulfillment and enjoyment in what he does.

In Jake's first ten years in the industry, he struggled with addiction, pushing himself to work long hours as a way to avoid facing personal challenges. His story is a reminder that, while this book emphasizes personal ownership, attitude, and mindset, we also need to recognize how overwhelming life can get.

Jake's journey shows that fun and fulfillment don't come from avoiding struggle, but from walking through it with people who've got your back. It's a powerful entry point into something we'll explore more deeply in the next chapter: how mental health fits into the way we lead and support one another.

Where We've Been, Where We're Going

After cutting into and peeking behind the soggy drywall with a flashlight, we've pinpointed the real cause of the leaks draining our enjoyment from our work. Sure, long hours and tight budgets play a role, but what really robs us of fun is a breakdown in the human element and the pipeline that connects us all—things that aren't in the specs but matter just as much.

Mental health struggles, trauma, and major life events don't stay at home when we come to work. That's why this human element pipeline matters so much. When we build real trust, show respect, and take the time to connect with our teammates, we create the kind of environment where people feel safe asking for help. That's how we make room for others to rise—not by being perfect, but by being present.

We understand the size and shape of the problem. But now the work has to start. In the next two chapters, we'll dig even more deeply into why these problems keep popping up, so we can stop patching and painting over them and start fixing them for good.

STEP THREE

ASSESSING AND PLANNING THE REPAIR

4

Owning Our Part: Standing by the Crew When Times Get Tough

"Sometimes the easiest way to solve a problem is to stop participating in the problem."

ANONYMOUS

ALL OF US working in this industry are part of something bigger than ourselves: a massive ecosystem that makes it possible to build the structures and systems that shape our communities, our society, and even the world. That's what makes construction so powerful.

But being part of this ecosystem comes with responsibility. How you choose to show up and participate affects more than just your own experience; it ripples through the whole system, for better or worse. And if your impact leans toward the worse, don't think it won't eventually boomerang back and hit you in the face.

Burnout

Around six years into my career as a project manager, the unrelenting pressure and mismanaged stress piled high enough to ignite early signs of burnout. The moment I got home, I'd pour myself a drink. I'd toss together anything resembling food, plop down in front of the TV, and let my brain zone out. Eventually, I would drag myself to bed—only to wake up at 3 a.m., haunted by project nightmares or my brain replaying the day and trying to solve every unresolved issue. Not to mention, I was probably still trying to metabolize the sugar from the alcohol I'd drunk earlier that night.

I never got nearly enough sleep before the morning alarm would go off. I'd pour some black coffee sludge into a travel mug and head back in to wade through the never-ending pile of problems.

Stress was past consuming my life. It was full-on digesting it.

Left unaddressed and festering, that stress began eating away at my health. My brain, my body, my ability to make clear proactive decisions—everything was deteriorating. I was short-tempered with my family and teammates. I started creating more problems than I was solving.

Using our drywall metaphor, I was painting over soggy drywall rather than acknowledging there was a leak. And when the paint started flaking and it started smelling of mildew, I blamed everything else: the humidity, the designer who spec'd the materials, the drywall finisher. Anyone but me.

I misidentified the "problems" as the main source of this stress. I deduced that if I eliminated these problems, it would eliminate my stress. So mission Eliminate Project Problems commenced.

It's Not Me, It's You . . .

Why should you care if other people are happy or having fun in the industry? You may be content as all get-out and having the time of your life with a smile plastered on your face, skipping onto the jobsite every morning.

Alternatively, you may be drowning in work, stress, pressure, and your own problems. So, screw everyone else, their feelings, and somber moods, right? We don't have the capacity to add more weight to our baggage and carry the burden and extra weight of anyone else's issues. It's easier to nod our heads and agree that we do indeed have a problem, and then blame other people for our problems, bad days, and lack of fun. We target incomplete drawings from the design team, leaky budgets from our pre-con team, a lack of communication and coordination between our office and field, supervisors and crew, or sprinkler fitters and tinners.

While deflecting or "gifting" these problems to others might feel like a relief, it's really just a Band-Aid on a boo-boo.

We can either invest our energy in planning ahead—taking steps to reduce how often we get cuts and scrapes, and learning how to properly respond when we do—or we can keep spending energy replacing bandages and risk the wound festering into something worse. Left unaddressed, it can infect everything—or consume a limb.

Here's the thing: if you are blaming anything, anyone, or everyone else, then *you* are the problem. Ouch, right?

You may or may not have been ready to hear this. It isn't an easy thing to recognize or admit to. Some of you may even disagree, close the book, throw it into the nearest Port-O-Let toilet, and walk away.

But before you do, allow me explain: every one of us in this industry has an impact on and is in turn affected by this "fun problem." We are all affected by this "leak." Admittedly, some of us more directly than others. Some of us are getting sprayed in the face every day, drenched in its contents while others are a bit futher away, and just sprinkled with foul-smelling remnants. Even those much further from it are still impacted by inhaling the mold particles from the drywall that slowly attack our respiratory health over time.

So, while most of us aren't actively trying to be the problem, many of us don't realize we are nevertheless participating in it. Every so often, even those who are actively trying to contribute to the solution are unintentionally the problem. We are human; we are imperfect.

The first step toward moving in the right direction is becoming aware of how we're showing up—and how this affects the ecosystem around us. Each of us has both the opportunity and the responsibility to recognize our impact, whether it's intentional or not. This awareness gives us the chance to stop contributing to the problem and start being part of the solution.

The Competent Person

Full disclosure: this chapter tackles some heavy topics, like substance abuse, depression, and suicide. If you're struggling with any of these, it doesn't mean you're "the problem."

Everyone has rough patches, and you deserve support, not blame. For the rest of us, we need to remember that how we treat folks on the job can literally save a life. We're in this together, and showing up for one another matters, even when the workload feels heavier than a wet stack of plywood.

At the same time, these aren't issues we can gloss over or patch with a quick fix. If we want to stop the leak—why construction no longer feels fun for so many—we have to be real about what's spilling out into our day-to-day. There's a cost to ignoring what's going on beneath the surface, and if we don't address it the mold just grows. Bottom line: for construction to feel fun again, we need to look these tough realities square in the face and take steps to support ourselves and one another.

I could have written this book to include nothing but glitter, blue skies, and plastic optimism, but this would be equivalent to putting a beach poster over our damaged drywall, misting some coconut-vanilla scent into the air, and saying, "See? All better!" But because I respect your time (and mine), and I care a lot about our industry, I refuse to do this.

This is not a "not my job, not my problem" kind of problem. This kind of problem impacts every single one of us working in this industry, which means that every single one of us has the opportunity and responsibility to participate in the solution. This is what will effectively fix the problem. This ownership is what will start improving not only our own outlook, well-being, and health, but that of our entire industry.

I understand some people get squeamish at the thought of taking responsibility and ownership over the unpleasant parts of life and our industry. Some would rather call a plumber to handle this type of work, rather than do it themselves, but here is the thing: it's our problem. It isn't anyone else's responsibility or job to tackle this. No one is coming to our rescue, and the solution cannot be outsourced or subcontracted.

Believe it or not, *you* are most qualified and competent to fix it because you know the affected people best. Your comrades, your colleagues, and your trade partners. You have built and hold the most trust and respect with these individuals. Relationships you have built with others in the industry over years and decades cannot be replaced. So don't underestimate yourself.

It's time to pull up our big-kid pants, put the glitter and air spritz back in the closet, grab our tools and work boots, and get to work on this leak.

Please know that mental health challenges and substance abuse are unfortunately common in our industry. You don't have to navigate these alone. At the end of this book, I've included a list of some of the resources available, with the phone numbers and websites of organizations that offer real support for those in construction wanting mental health support. You don't have to navigate this alone.

Construction Is a Swamp—I Mean, Ecosystem

Proliferating Organisms, Overinflated Egos, and Eroding Self-Awareness

Our industry is like a swamp. While we've all heard the sandbox metaphor, it doesn't do us justice. While this swamp analogy may seem disgusting, or possibly even insulting, allow me to clarify.

Swamps have depth. They are incredibly important ecosystems, providing essential nutrients and habitats for everything in and around them. As an environment, swamps host countless species of life: plant, animal, fungus, and bacteria. They're vital parts of our world, cleaning the water that flows through them, reducing flooding, and returning nitrogen to the atmosphere. Just as construction is critical to society and communities, swamps are critical to our environment and world. We need construction. We need swamps.

What we don't need is a chlorinated swimming pool. While it might look cleaner, it's dead and artificial. If our industry were like a swimming pool, it would be not only boring, it would be filled with suit-wearing, pretentious, egotistical individuals. This isn't to say that everyone in a chlorinated pool industry environment is a shallow snob, but most of us in the construction industry would rather lose a finger to a table saw than work in this type of sterile environment. For most of us, having to wade through and navigate such a prescriptive and unchallenging environment would be torture in itself. We are much more gritty and interesting than that, for goodness' sake.

The dense swamp is more complementary to our wholesome, abundant industry. It better represents the type of complex problems that challenge us. We—the individuals working in the industry—are simply living organisms of this swamp, challenged with trying to tread, navigate, and swim through it. Not everyone is cut out for the challenges our industry doles out. It's not easy, and it takes intestinal fortitude and a relentless type of gumption that not everyone enjoys or is even capable of.

The construction industry is an amalgamation of all of us working in it, in the same way a swamp is the sum of all the organisms. This includes those in the field, putting physical work in place. It includes all those who support the field in management, accounting, marketing, design, safety, etc. It encompasses those working to ensure the correct materials are available and delivered, and that any obstacles are removed to give the field a clear runway to put work in place. It requires all of us, from varying backgrounds, skill levels, personalities, and expertise, to keep the ecosystem in balance. We may be the ones swimming through this swamp, but we also influence its condition—every day. The filtered water, the clean air, the shelter—these come from the way we work, lead, and show up for one another. We don't just exist within this environment; we help shape it. For example, if one organism proliferates and dominates in a swamp, it may kill off other critical organisms, and the swamp may then stagnate. The same goes for our industry. We are a team: our experiences, actions, reactions, and decisions impact those around us.

This means we both influence and are influenced by the people we work with and alongside. Every one of us contributes to the reputation of our industry and the environment we operate in. Whether our impact is good or bad, intentional or not, we're always participating. And at any given moment, we're either contributing to the problem or helping build the solution.

Gift-Wrapping Problems

As a project manager, I was always the most important person on a construction project. The project wouldn't happen without me. I was the one who negotiated and drafted all the subcontracts. I got them all on board and dictated where they resided within the master schedule I had created. They wouldn't get paid without me approving their change orders or invoices. Yeah, sure, the actual work wouldn't *physically* get done without those in the field (and everyone else), but they couldn't do this without a promise for payment and the oh-so-motivational "back charge" or "notice of delay" emails every so often, right?

This is, of course, ridiculous and untrue, although that didn't prevent me from thinking it when I was green in my career and obviously had a bit of an ego. While my role as the project manager may have meant I had a lot of responsibilities, it's arrogant to claim I was the most important player. In fact, my role wasn't ever entirely critical or essential. I was very helpful, though, to our critical and essential field folks. I was there to support them—to ensure they had decisions and materials out in front of them to put actual work in place. I removed red tape and barriers and reduced risk to allow them a clear runway.

My ego was so inflated in those early days that it blocked my view of seeing all the damage I was causing to my projects. Here I thought I was leading and helping the team, when in reality I was bulldozing relationships and creating rifts. That critical human element pipeline—the communication, trust, and respect—started to break down and weaken the entire team. The energy and focus required for doing the actual project shifted from supporting the team and became more about competition, about who was right, who was in charge, and who was at fault for any of the project's problems.

Another unfortunate side effect of this arrogant approach of mine was that I unintentionally created more problems for myself and my teams. I determined that other people were the cause of my problems and stress. This was likely due to a fear of failure or of exposing my own insecurities.

I avoided, passed off, or tried "gifting" these project problems to others. I also did things *my* way because I felt our industry's outdated processes were stupid, and I knew how to do them better. And being the most important person on the project, *the others* were the ones who should change and should all mold to *my* way of thinking. So, in a poor attempt to reduce or alleviate my own stress and reduce my problem pile, I blamed others to try to shift this inconvenient weight off my shoulders (and onto someone else's). I figured then construction would be fun again, at least for me.

You may have experienced something similar within your career. Again, you might not have done it intentionally, but this lack of awareness comes at the cost of everyone else *and* yourself. In trying to bury yourself in the sand to hide from problems, you end up kicking up silt and mud, making it murkier for everyone else, too.

Overinflated Egos

Overinflated egos are a major contributing factor to our industry's fun leak. They create a false sense of competence, authority, and importance, making us believe we contribute more than we actually do, and they often lead to micromanaging. An inflated ego can feel like some supernatural power that makes us think we can control and manipulate other people or things we have no right meddling with, but in reality it's not power at all—it's poison. It's a false, hollow sense of strength. The more inflated our ego becomes, the stronger we think we are, but it's actually a crippling weakness, blinding us to the very problems we are causing. The more inflated the ego, the weaker we become, trapped by our own delusions of control. Worse yet, it blinds us to opportunities for genuine growth and problem-solving because we refuse to recognize that the problem exists.

I've learned firsthand that bringing an unchecked ego into construction is like stepping into a professional boxing match with inflatable gloves and no strength training. You may feel prepared. You may even last a round or two. But ultimately, you'll be knocked out by those with real skill, real discipline, and real teamwork. Just because you're unaware of your ego doesn't mean others don't see it—or that it doesn't impact them. In a high-stakes, collaborative environment like construction, an overinflated ego is not just a personal flaw; it's a team killer. It weakens communication, erodes trust, and slows problem-solving, forcing others to navigate unnecessary conflicts rather than focusing on the job at hand. If you don't manage your ego, your teammates will have to manage the fallout—and that's a guaranteed way to weaken the entire project.

The Submittal Boxing Match

One of the major sources of my project problems early in my career was submittals. They were (according to me) frequently incomplete and improperly submitted, with materials incorrectly substituted. This created more work for me. And more responsibility to catch and correct the errors or omissions—and, consequently, stress. To reduce this "problem" and reduce my workload, I figured I would instruct all the subcontractors on how to properly compile and submit these documents. Surely, they would have compassion for little ol' me and the amount of work I had to do? This would then motivate them to put more energy into ensuring their submittals were done perfectly every time.

My attempt to solve this issue included spending approximately three hours crafting a two-page Word document titled "How to Submit Submittals Properly." This would not only teach them the importance of submittals, but also outline every step in how to verify their own materials before sending them to me. If they took the risk of bidding an alternative material, they needed to send me a comprehensive breakdown of the differences in their materials and list how this benefitted the client. Happy with the extensive detail in which all this was explained, I filled the "To" line with the email addresses of all the subcontractors on one of my upcoming projects and confidently pressed "Send."

I'm sure you can imagine how this went over. My inbox was silent. I'd likely irritated my trade partners with my all-important, woe-is-me attitude. My 27-year-old self didn't really think through the consequences of that one.

Luckily, I didn't get any snarky or snappy responses—at least, none that ever came back to me. One subcontractor approached me after our big project kickoff meeting, and kindly and respectfully informed me that he had read it and advised me that I should consider retracting it. It was the kindest way to smack me over the head and say, "Nice work, dummy!"

The lack of response and his wise advice helped me realize that my message didn't just miss the mark; it ticked people off and shut them down. I hadn't earned trust. I'd bulldozed right over it. Great. Now I had an entire project in front of me with all these skilled professionals who had just witnessed me trying to pass off and shirk responsibility. It took a lot of digging to get out of the hole I'd created for myself, but I did finally earn back the respect and trust of *most* (but not all) of my subcontractors.

While this attempt was an embarrassing failure, it came with valuable lessons—ones that have become permanent fixtures in my memory. They hang in my mind like those sarcastic motivational office posters: always there, always reminding me.

This one in particular taught me that isolating yourself and leaving your teammates in the dust doesn't make you a leader, it just fast-tracks your downfall. Throwing others under the bus doesn't earn respect or results; it just leaves angry people in your wake. Spending three hours trying to dictate and micromanage everyone wasn't just a waste of time—it made the job harder. Petty mistakes like this create more work in the long run, especially when it comes to repairing relationships that didn't need to be damaged in the first place.

These kinds of failures still echo in my mind, but the deeper, heavier lessons? Those are the ones etched into my gut.

Consequences: Pay Me Now, or Pay Me Later

In 2007, I left the second of my three construction employers. The reason? Again, my lack of self-awareness and maturity to take responsibility for my decisions and reactions. While this isn't what I explained to my supervisor as to why I was leaving, they embarrassingly saw right through it and commented on it during my exit interview. While I didn't care to admit it at the time, I allowed a fellow project engineer's competitive and seemingly brown-nosing tactics to get under my skin, and he was the main reason for my departure.

Jarrod* (as we will call him) didn't even interact with me all that much. While we both worked out of the jobsite trailer, we worked on separate mechanical scopes of work on this particular project. Yet, for some reason, I decided he was the cause of all my problems, and unfortunately he became the target of blame and of my sour behavior. I didn't particularly agree with his style of management, work, and communication methods. I decided this was irritating, and allowed it to consume my focus.

I found myself competing with him. Competing for what, I don't know, but it became so concentrated that it started leaking out in my interactions with other team members.

"Well *Jarrod* didn't do this, so I had to," I'd say. "I should go to the boss first to let him know it was my idea before *Jarrod* takes credit for it, *again*." I thought I was containing and hiding this animosity (or maybe jealousy) well enough. I was professional when appropriate, and held my eye rolls until I was out of sight. But one day I slipped, and my resentment resulted in an accidental "Reply All" scenario that Jarrod was cc'd on, which included a sarcastic comment about Jarrod.

The moment I pressed "Send," I had one of those *oh crap* moments that feel like you've just been punched in the stomach. Then the sense of dread and shame comes over you as you recognize your mistake and your embarrassing behavior. It was not a proud moment. At this time, I apparently decided not to take ownership and responsibility for my mistake. Instead of walking over to him in the trailer and apologizing, I blamed it on some ridiculous circumstance, twisting the intent and meaning so that I wouldn't look bad. I don't think he bought it. He didn't retaliate, or call me out, but it was quite obvious he was not interested in my excuses. He simply nodded, looked back at his computer, and continued working.

Luckily the comment wasn't anything too scathing or incriminating that would get someone sent to HR. I felt I had quickly swept it under the rug, and I moved on.

In my mind, I framed it as another important lesson learned. Even if I don't agree with or particularly get along with some teammates, every person brings their own value to the team. I also learned that certain things will always fall within my control, like my behavior, my communication, and how I choose to show up.

And maybe most importantly: always double-, if not triple-check the "To" and "CC" lines before hitting "Send." Better yet, don't waste your energy on petty jealousy or gossip in the first place.

While those takeaways did improve how I interacted with people—even after I left that company—I didn't realize until much later that what I thought I had swept under the rug would eventually come back. And when it did, I had no choice but to take ownership of it.

A few years ago, I happened to run into a colleague who had worked with us on that project. During our brief conversation, he informed me he'd heard Jarrod had recently died by suicide.

Talk about a gut check. I realized I was not directly responsible for this tragic situation as our interactions were so long ago and as insignificant as a piece of hay. But was I not, at least indirectly or the smallest amount, part of the problem? A piece of hay isn't heavy, but a bale of hay is very heavy. If Jarrod's collective interactions with others had been similar to mine, each adding their own piece of hay to the pile, it could very easily have become an unmanageable crushing weight. Because of my own selfish insecurities and inability to take ownership of my actions at the time, I can tell you for certain I wasn't a part of the solution.

This is forever my guilt and burden to carry. I cannot apologize, I cannot clarify, I cannot relieve him of this, bring him back, or help him now in any way. His wife no longer has a husband, his kids no longer have a father, his parents no longer a son, and our industry no longer a valuable and much-needed teammate. That small piece of straw that I added to his burden, that I refused to admit and take ownership over at the time, has now been exchanged for a heavy rock that will forever be stored as an important lesson in my gut. This heavy lesson is now mine to carry with me for the rest of my life.

As one industry member so eloquently explained one day over lunch: "Construction is a 'pay me now, or pay me later' industry. If you aren't doing what you are supposed to be doing now, you will pay later, and it won't be fun."

Boy, do I wish I had been given (and accepted) this advice earlier in my career. These lessons will claim your ownership of them one way or another. Either in the moment, when you have more control and opportunity to participate in the solution, or later, after you have tried to avoid, bury, or deny it—when you have less opportunity to fix it, and its consequences have become a much larger and heavier burden to carry.

Luckily, at some point along my career, I finally got my act together and matured out of this arrogant, egotistical, and selfish way of thinking. I caught a glimpse through a wider lens and saw our awe-inspiring industry for what it truly is and all that it offers, rather than what I thought it should be and all the barriers I felt it was throwing in my way.

I recognized just how naïve and shortsighted my approaches were in trying to change other people into something I thought would benefit *me*, instead of appreciating who they are and all they already offer and provide. I had to stop trying to change our industry into something I thought it should be. Even before taking the time to recognize and appreciate what it already is.

It wouldn't be as easy as deciding to go to work tomorrow and seeing everything as peachy and fun. We aren't miracle workers, but I could start doing something about it. If I wasn't happy, if I wasn't having fun, if I wasn't proud of my work, trajectory, or career, then sitting around whining, blaming, and expecting everyone else to do something about it wasn't going to get me anywhere but further down the hole I was digging for myself.

I will tell you, admitting that my lack of fun, fulfillment, and enjoyment was my fault, and that I was the problem, was not a fun, easy, or quick process. Neither is taking ownership and responsibility over the consequences. Even after that "ah-ha" moment, it still took many years to kick the bad habit of trying to blame my lack of fun and enjoyment on others or the industry. The moment you finally realize you have complete control over this—your experience, your destiny, and impact—then it becomes one fantastic, energizing, and freeing feeling.

The Risk of Waiting or Doing Nothing

Unfortunately, Jarrod isn't the only co-worker I've known who has died by suicide. And it wasn't until 2018 that I finally came across some hard facts: that most of us in this industry have probably lost someone we know—a teammate, a buddy, a co-worker—to suicide. Construction has the second-highest suicide rate of any occupational group in the US, behind the oil and gas industry.[21]

It's not just here in America. One study found: "A quarter of the construction employees in the UK have considered taking their own lives."[22] That's not a small number. This is a substantial problem our industry faces.

Here's the kicker, though: we've only started talking about this recently. It's not that this problem suddenly appeared in

21 Cora Peterson et al., "Suicide Rates by Industry and Occupation — National Violent Death Reporting System, 32 States, 2016," *Morbidity and Mortality Weekly Report* 69, no. 3 (2020): 57–62, https://www.cdc.gov/mmwr/volumes/69/wr/mm6903a1.htm.

22 Daisie Rees-Evans, *Understanding Mental Health in the Built Environment (Chartered Institute of Building*, 2020), 4, https://www.ciob.org/industry/research/Understanding-Mental-Health-Built-Environment.

the last decade—it's just when researchers finally decided to focus on it.

Cal Beyer has been on the front lines of this for years. He's one of the guys in our corner when it comes to mental health, suicide prevention, and substance abuse in construction. "I've been doing safety and risk management since the '90s," Cal told me once. "And early on I had a mentor, Bob VandePol, who taught me the basics of psychological first aid. That changed how I thought about all this."

Like Cal said: "There was no real data on suicide rates, sorted by industry, until 2016. Even then, the study got pulled and rewritten in 2018, and even the newest report we have is still using numbers from back in 2021."

At one point, Cal even started working toward a doctorate just to figure out how to help our industry. He soon realized that the information he needed simply wasn't available yet. Rather than waiting for perfect data, he chose to act. "We didn't have time to sit around waiting," he said—and he was right. So he rolled up his sleeves and got to work.

He started writing and speaking up. His first article was about what he called "the missing human element" in crisis management—basically, how to take care of people when things go sideways. But it wasn't easy getting folks in construction to open up about mental health. "For years, nobody wanted to talk about suicide," Cal said. "It took a long time to tear down the wall."

By 2010, he had helped start a workplace task force for the National Action Alliance for Suicide Prevention. In 2012, with the help of his mentors, he kicked off a construction-focused subcommittee. Since then, he's written over 150 articles and given hundreds of talks all over the country to raise awareness and push for change.

Cal has done a lot of valuable work, but there's something he said that stuck with me in particular: "This isn't just about mental health. It's about all the stuff we're carrying around. Problems at home. Legal issues. Money worries. It all piles up." And one of the biggest things weighing people down? Substance abuse.

"Before the 2015 SAMHSA [Substance Abuse and Mental Health Services Administration] report," he told me, "nobody really knew how bad it was. That report showed construction had the second-highest rate of substance misuse, right behind mining. And even with that information, it was tough getting people to talk about it."

"Everyone was fine with drug testing and zero-tolerance policies," Cal said, "but no one wanted to talk about prevention or recovery."

Then the opioid crisis hit, and it hit us hard. "I saw it coming," Cal said. "We've got some of the highest rates of injuries out there—back, knees, shoulders, you name it—and that means prescriptions for pain. But it doesn't stop with the worker. Sometimes those pills end up in the hands of kids or family members. That's how addiction starts. That's how overdoses happen."

It wasn't until the COVID-19 pandemic that we finally had to face the music. In 2023, the Centers for Disease Control and Prevention (CDC) put out overdose death rates by industry—and guess who was number one? Construction. "Once that report came out," Cal said, "people couldn't ignore it anymore. That's when things finally started moving in the right direction."

By February 2025, at the Fourth Annual Construction Working Minds Conference, Cal was leading a half-day program on overdose prevention. He delivered the keynote, facilitated recovery strategy panels, and ran breakout sessions on best practices. "We're finally looking holistically at all the major risk factors," he said. "And it's long overdue."

While this information has been distributed across social media, spoken about and shared in industry conferences, and plastered over industry publications, there are still many in our industry who are unaware of these significant statistics. It's been a major limiting factor in igniting urgency on these issues.

The Skilled Labor Shortage: The Other Big Fight

Even though society is still wholly dependent on us to build and maintain the very structures that sustain modern life, rarely are we recognized and appreciated. On top of the above statistics, our industry also has a systemic labor shortage, despite substantial recruitment efforts. As mentioned earlier in this book, movies, media, and entertainment (sadly some of society's main education sources) have previously portrayed us as hillbilly dropouts who are just out to swindle them out of their hard-earned cash by way of change orders. This tainted perception of who we are and all we do in construction continually works against us in our struggle with labor shortages—i.e., people not choosing to enter this industry or actively choosing to exit.

I have never had the privilege to experience our industry without a labor shortage. I don't know of anyone in our industry who has. Between this societal misperception of who we are and the very real problems of stress, burnout, suicide rates, substance abuse rates, and mental and physical demands, we are fighting an uphill battle. With fewer people entering the industry and many others leaving for various reasons, it feels like trying to use a residential water hose to fill a five-gallon bucket that has a two-inch hole in the bottom.

This has meant we've had to get more done with less people. To do this, we over-task and overload our current labor force. We pile on more expectations, responsibility, workload, and hours. Which subsequently means extra micromanagement, pressure, and the increased likelihood of burnout or injury.

While our industry's labor shortages, deaths by suicide, and substance abuse problems have been verified and brought to the surface by researchers, our industry's stress levels, divorce rates, and health problems (mental and physical) are not as easily measurable or quantifiable. This isn't to say we shouldn't research these or that researchers aren't trying with all their might—and as quickly as possible—to help bring awareness to these issues. Unfortunately, this isn't an easy feat or a quick process.

As you and I know, our industry isn't one to enthusiastically drop whatever we are doing when a PhD student walks onsite wanting to study us. Many of us don't identify as enthusiastic survey takers either. This makes it exceedingly challenging for researchers to get their hands on this critical information. On top of this, if and when researchers are able to mine this data from our industry, it can take years to sort through, analyze, type up, get reviewed, and then get published and back into the hands of our industry. No matter its value and importance, we can't afford to wait until research catches up.

Ink Versus Instinct

Unfortunately, all of this means I cannot provide a stack of peer-reviewed research confirming that fun, fulfillment, and enjoyment are directly, or even indirectly, related to our industry's current hemorrhaging problems. I could choose to invest the next twenty years researching and verifying this so I could then officially say, "See? I told you so!" Any satisfaction derived from being right, however, would be greatly overshadowed by the damage caused by not proposing solutions and doing something about what I know, down to my core, to be true. Even more lives will have been lost, many industry colleagues worn down, their health taking the hit, and many others will have left the industry—fed up or burned out by then. It's a matter of ink versus instinct, and we need to trust our instinct until the ink catches up.

This isn't to deter or diminish any of the current researchers tirelessly working to provide this research. I have and will continue to support and encourage those working toward proving these facts and providing this critical data. But as with some of the decisions on our projects, we can't sit around and always wait for one hundred percent confirmation and the paperwork to catch up before we have to make a decision, move forward, and do something about what we instinctively know to be the problem and solution. We are the ones living, breathing, sweating, crying, and bleeding this industry every day. All of us living this lifestyle are impacted. We have either experienced one of these struggles ourselves or have close workmates who are struggling with at least one, if not many, of these.

Wider Industry Lens

This lack of labor, support, and help—combined with demanding expectations from society, clients, and teammates—means we can sometimes feel like we are supposed to be everything to everyone. The master of all trades.

Superhuman. This is not only impossible but also exhausting, and this can sometimes feel like we're running on a hamster wheel.

We are supposed to work long hours, but also get enough rest and sleep to recharge, so we can show up the next day with physical energy, sharp problem-solving, and sound decision-making. Yet when we finally do fall asleep, we often find ourselves wide awake at three in the morning, our brains racing to solve problems and churn through our endless to-do lists. Stacked expectations lead to bad sleep, which may be exactly why some of us turn to mind- and body-numbing substances after work in a desperate attempt to cope with these superhuman demands. Then, the next morning, we rely on a pot of coffee or a pack of energy drinks to wake up, push through, and survive another day. All the while, we tell ourselves that because we're working hard and consistently moving our part of the project forward, we're contributing to the solution—not the problem.

But when we live this way, we burn ourselves out just to get the job done. We drain every ounce of energy throughout the workday, leaving almost nothing for our families, friends, or hobbies, let alone the energy to make good decisions at work the next day.

So how do we get off this bloody merry-go-round? How the heck are we, as individuals, supposed to do something about it? If not for ourselves, at least for our teammates?

Originally, I thought that fixing this leak would require a massive intervention, a substantial amount of effort or work, or would take some miracle drug to make our industry fun again. That the entire industry—billions of people and hundreds of years of momentum—had to change in order to flip the override switch.

Luckily, considering our leak narrows down to the human element, the solution is much simpler than this. Not necessarily easy, but simple. And it is something that each of us, individually, can do.

You have full control and the capability to make construction fun, fulfilling, and enjoyable (again), not only for yourself but also for those you work with and around.

Snowball Effect: Your Choice, Your Control, Your Impact

My submittal micromanagement, my gifting of blame, and the email blunder I mentioned earlier are actions, behaviors, and interactions that take place in our industry daily, on projects across the world. While they may seem inconsequential or minor at the time, you can see the unintended snowball effect they can have. This is true not only for myself but for many others, as well.

I was trying to control things I never actually had (or will ever have) control over. And ironically, I lacked awareness of the one thing I did have control over: myself. I failed to consider interactions, behavior, and reactions. I saw certain aspects of our industry (and my tasks) as burdens; they were problems, rather than potential opportunities, gifts, and rewards. All of this was within my control, yet I didn't choose to invest my energy and focus on this. Instead, I spent my energy on inflating my ego and building armor to hide my insecurities. It was exhausting, worthless, and ultimately embarrassing.

If you look for the negatives, you will find them.

If you look for the positives, you will find them.

The times I considered myself the most important person (MIP) on a project were the times I was the problem. Or at least my ego and lack of self-awareness were the problem. The very problems I was trying to avoid, reduce, or eliminate were the same problems I created and was contributing to (and complaining about).

I may not always like everyone on my team. While it is not and cannot be my personal responsibility to ensure everyone else is happy and enjoying their careers, or making it easier for everyone else, it is at least my responsibility to ensure I am not making it harder or less enjoyable to be a part of the team. Ultimately, it was my choice to see Jarrod's actions as annoying and brown-nosing. It was my choice to allow it to get under my skin. All of this was a major reason (excuse) I decided to leave the company, my project, and my team—the ones who needed Jarrod's help *and* my help to carry the load and complete the project tasks.

At the time, I convinced myself that I was leaving them with this extra load because the company and management were immature and childish. Ironically, it was my immaturity and lack of emotional intelligence that caused me to leave. It was my own mindset, decisions, and actions. Not the company's, not Jarrod's, not my team's, or anyone else's. All of it—the email, how I responded to my mistake, deciding that Jarrod's personality and style were annoying, and allowing it to consume my attention—was within my control, and my reaction was all my choice.

It was also my choice to see submittals as a burden rather than the opportunity it was. I labeled it a problem task that caused me stress. This mindset contributed to my interactions with my trade partners and my teammates coming off in a way that was controlling and selfish. I was trying to reduce my workload by adding more to theirs. Simply because I felt I was the most important team member on a project—that I was somehow above doing these piddly project tasks.

This narrow-minded approach not only lost me the respect and trust of my valued team members, it also blinded me to the value, rewards, and opportunities right in front of my face. These tasks are critical to the smooth operation of the project and are also some of the richest learning opportunities. Submittals provide the opportunity to dig into every detail and crevice of the drawings, allowing you to be the go-to resource for questions. RFIs provide the opportunity to gain perspectives and language from trade partners to translate between designers, clients, and tradespeople. Meeting minutes provide phenomenal translations of the team's interactions, commitments, and challenges, as well as being a historical document of the entire project and project team. I failed to recognize these as rewards, as opportunities to continually build skills and habits to become one heck of a construction leader! Instead, I chose to focus on the negatives. I chose to see them all as barriers to my success and threats to my ego.

Ultimately it was my choice and within my control to experience the industry as either enjoyable or irritating. And my choice directly impacted whether I was a part of the problem or part of the solution.

Construction is a team sport. There is no one person in our industry or on a construction project who is the most important. The MIP, so to speak, changes on any given day. Our construction projects cannot be developed, designed, managed, and built by one person. It requires a team: a balanced, functional ecosystem, with varying backgrounds, skills, energy levels, perspectives, and support to build these structures and systems that keep society functioning and operational.

I didn't know it at the time, but those failures were amending the soils and laying the foundation for something more important: repair. It's one thing to talk about accountability and having the guts to face the truth; it's another to live it when the damage is already done. The real test comes after the dust settles—when you have to circle back, own what happened, and start rebuilding what you broke. It means recognizing when you're part of the problem, admitting it, and actively working toward solutions. That's where we're headed next.

5

As-Builts: Doing Something About It

"Whether you think you can, or you think you can't—you're right."

HENRY FORD

AS HUMANS, connecting with others isn't just a choice, it's a primal survival need.

Our brains and bodies are wired to seek out and remain a part of teams, crews, and communities. This means working together, building relationships, and having each other's backs—all of which forge that human-element connection.

But these connections don't happen without effort or intent. Our brains only create those feel-good rewards—like fulfillment, pride, and respect—when we put in the effort. Fun isn't just given to us; it's built through intentional quality interactions—by showing up and putting in the work. We only get better at what we practice. So let's start working on it.

We've found the leak, and now it's time to understand the full scope of the damage. It runs along the main pipeline buried deep within the human element. One side of this pipeline represents our construction work environment; the other side holds the sense of fun, fulfillment, and enjoyment we hope to extract from it.

This chapter takes a closer look at some of the human element specs, submittals, and as-builts—so to speak—to show how we can reignite and engage fun within our work.

It's important to clarify that, while I am here to support you, I cannot do this for you. No one else can either. But I believe in you. If you've worked in our industry for any reasonable length of time and are still working in it, then you inherently have the skills, intelligence, and ability to tackle this.

So let's start inspecting this pipeline, shall we?

Work Boot Scenario

No matter your role, I'd like you to mentally put on your work boots, and imagine you work in our industry's vertical commercial sector as, say, a mechanical foreman. You've worked your way up to this role over several years and now lead a strong, competent crew of pipefitters. You are assigned a large project: a new ground-up building for a national semiconductor producer. Let's say the mechanical contract alone is around $18 million, contracted under the GC.

Your most recent project was one of the most challenging projects of your career, with incomplete drawings, outdated specs, disjointed collaboration, and hundreds of avoidable RFIs driving cost overruns and safety incidents. Although it finally reached substantial completion, you're still dealing with warranty requests and closeout items, but you're eager to start fresh on the next project with the same GC, albeit a with new team (i.e., superintendent, foreman, and project manager).

It's a cold Tuesday morning, and you're headed to the jobsite for the first time to attend the weekly subcontractor meeting held in the GC's jobsite trailer. You've arrived on site 15 minutes early, noticing a handful of vehicles and workers putting on their work boots and personal protective equipment (PPE) and collecting their tools from their trunk or truck beds. You throw on your safety vest, grab your hard hat (or helmet), your safety glasses, a rolled-up copy of the project's site plan with your sketched location for your trailer, an aluminum clipboard holding your red-lined contract, and a notepad and pencil, and head to the main trailer.

You walk up the aluminum steps, kick the loose gravel and mud from your boots, open the door, and notice no one is in the trailer yet. You take in the makeshift offices along the trailer perimeter. Compliance posters and a large first-aid kit hang on the fake wood paneling. In the center are two large foldable tables, pushed together, with chairs around them. You settle near the far end, facing the entrance.

Six minutes pass before you hear footsteps on the trailer steps. Someone opens the door, holding onto the doorknob while kicking off their boots; he is followed by two other people he is engaged in conversation with. They continue discussing and debating among themselves as they take their seats in the chairs closest to the door, catty-corner from you. Thirty seconds later, another person enters the trailer. A hard hat is tucked under his arm, and he has a notebook in one hand and his phone held up to his ear in the other. He seems to be finishing a conversation with what sounds like someone from his office on the other end. He completes his discussion, then takes a seat perpendicular to you at the end of the table, near the door. He nods his head at you in greeting as he sits, then proceeds to type something on his phone—likely an email or text in follow-up to his discussion.

The trailer door opens again two minutes later, and a handful of people walk in. Two of them sit in the same row as you, but still a few seats away, and pull out their phones. The other three remain standing back behind phone-conversation guy, even though there are still a handful of empty seats. Two of them have their hands in their pockets, and the other leans against the wall and crosses his arms.

Moments later a younger gentleman, dressed in clean slacks, loafers, and an ironed, button-up, collared shirt, enters the trailer. He isn't carrying anything but his phone, which he seems to be intently focused on. He sits next to you, hunches over in his seat with his eyes plastered to whatever he is scrolling through. No one else seems bothered or even acknowledges his entrance or behavior.

Thirty seconds before the start of the meeting, the door opens and in walks a man wearing nice jeans; an ironed, collared shirt; and a Carhartt vest covered with a safety vest. As he is kicking off the mud from his Kodiak work boots, you notice his full-brim hard hat—heavily scuffed, but clearly cared for. When he looks up to walk in the room, you spot label maker tape stuck beneath the GC company logo with a name you can't quite make out. He makes eye contact with the three guys standing up in the back and shakes the hand and pats the shoulder of one of them. He turns and asks the three people opposite you if they "got it all worked out," to which one says, "Not yet," but there seems to be air of confidence that it will get resolved. He makes his way over to a seat perpendicular to you on your right, at the head of the table, furthest from the door but facing it. He sets down his phone and an aluminum clipboard that is very similar to yours. He takes out a pen from his safety vest pocket and puts it down next to the clipboard. Remaining standing, he looks over at you, offers a handshake, and introduces himself as Matt. You stand, introduce yourself, and then both of you sit and pull in your chairs.

Everyone in the room quiets down and gives their full attention—except for the gentleman next to you, who's still thumbing through something on his phone under the table. A few people pull out hard copies of what appear to be the agenda, while the man on your right switches on a large TV screen behind you that projects the agenda to the room.

You're momentarily distracted by the sound of heavy, deliberate footsteps on the aluminum stairs. A few people in the room shift in their chairs or sit up straighter, though their eyes stay fixed on the screen. The door opens, letting in a quick blast of cold air and trailer dust. A woman steps in, wearing boots (not quite work boots), slacks, and an ironed button-up shirt topped with a safety vest that appears to be slightly too big for her. She knocks extra mud from her soles at the threshold, then walks in, her movement confident but careful on the vinyl flooring, still marked with half-dried boot prints from earlier arrivals. She heads toward the seat directly across from you.

Her wide-brim hard hat, stamped with the GC logo, is clean and still has a bit of shine, and the name "Amy" is printed in block letters on a strip of label maker tape stuck beneath the logo. (Yes, this is me.) She makes eye contact with you and offers a quick smile and head nod. You notice the man to your immediate left finally puts away his phone as she hands him a neatly clipped agenda packet. She thumps a thick stack of papers onto the table, pulls out a chair, sits down, opens a weathered portfolio binder, and then looks to Matt, greeting him warmly but respectfully. He returns the greeting, and the meeting begins.

While there seems to have been very little verbal communication or interaction happening in this jobsite trailer over the past fifteen minutes or so, there were actually bucketloads of nonverbal signals hitting all our senses. Even if we don't realize it, we're constantly absorbing and processing these quiet cues—signals that shape how we size people up, how we engage (or don't), and how we operate moving forward on the job.

What messages were flying around that room? Could you gauge the role or experience level of the guy glued to his phone? What story did you tell yourself about the ones standing in the back with their arms crossed? Or the clean, polished guy who came in empty-handed? How did Matt's entrance—or mine—shift the tone of the room?

We make split-second judgments constantly. This isn't a flaw; it's a survival instinct. Our brains are wired to scan for status, threat, and credibility, finding and creating patterns to help us thrive. And where does this instinct come from? It's built on everything we've been through—our past experiences, the crews we've worked with (and against), the jobs we've run, the mistakes, the wins, and the wisdom we've earned along the way. All of this provides context for the silent language being spoken and the messages being sent. It's not something to ignore—it's something to value.

But like any tool, it only works if we know how to use it. The key isn't to stop making snap calls—it's being aware that we're making them, and understanding how and why they happen. That's how we figure out when those instinctual reactions are helping us versus when they're getting in the way.

Human Element Pipeline As-Builts

Not being able to see or touch this buried human element—and its "connection pipeline"—can be difficult. Yet in many scenarios, such as stepping into a trailer for a meeting, each of us senses something slightly different about the room's mood or the group's dynamic. Many of us can gauge the tone and pace of a project before anyone speaks, because our brains are exceptional at processing information. In fact, we take in more information from our environment—what we see, smell, taste, touch, hear, and sense in other ways—in one day than the largest computer does in a year.[23] We filter this influx of data, discarding what we consider garbage and focus on what seems worth keeping, sending it through our pipeline for further processing. The content then gets mixed, creating concoctions that formulate meaning and value to generate the actions, reactions, and interactions that shape our environment.

To help visualize this hidden process, imagine your "human element pipeline" as a loop running in and out of a building—representing your brain and body—with an inlet gate on the outside. This inlet gate is your senses, constantly drawing in data from the environment. In front of it sits a

---

23 David A. Sousa, *How the Brain Learns*, 5th ed. (Thousand Oaks: Corwin, 2017), 46.

"trash grate" catching some of the irrelevant or overwhelming details while still letting plenty of information pass through.

Once inside, the content flows through your chosen mindset filter and energy gauge and, along with your past experiences and core values, interacts with potent mixes of hormones and neurotransmitters. This mental "treatment plant" shapes how you interpret what you see, hear, smell, taste, touch, or otherwise sense.

Finally, it sends this customized blend of emotions, decisions, and behaviors back into the environment—like water flowing outside—where others can absorb it or where it can be sucked back into your pipeline again. Despite the complexity of this system, it all happens so seamlessly that we often forget or don't even notice how much information we're actually taking in and processing every single day.

This overview serves as the "site plan" for the system—a broad architectural view. For a deeper dive into the underlying mechanics—much like examining the cuts, elevations, and section details of a blueprint—I encourage you to check out Appendix 1, where we explore these pipelines in a bit more detail.

Think of the meetings and jobsites you've been on. The moment you pull onto a jobsite, every unspoken detail—from the order (or chaos) of the materials to the distinct smells, sounds, and body language of your crew—sets the tone for the day and the project. Your senses are busy taking in a flood of information, feeding into your personal "human element pipeline" that influences how you feel, decide, and act.

Now, we're starting to pinpoint where our control lies within this intricate system. Somewhere along this main pipeline, where our tough construction work meets the fun, fulfillment, and enjoyment we all crave, there's a "leak" waiting to be fixed. Again, let's dive a bit deeper into this blueprint and uncover how you can control your input and outputs.

How Fulfillment Actually Works: SDS for the Human Element

Before we get into brain chemistry and mindset, I want to be clear: I'm not a doctor, therapist, or neuroscientist. What you're about to read isn't medical advice. It's a collection of ideas pulled from research, personal experience, and a desire to make some complex stuff feel relevant and useful for those of us in construction.

When I talk about hormones, neurotransmitters, stress, or mindset, I'm not claiming to cover the full picture—or suggest this is the only way to look at it. Science evolves. There's room for many views. My goal here is to offer a perspective that helps you better understand how you show up and what's within your control, your influence, or neither.

If any of this catches your interest, awesome. Keep going. Read a book. Listen to a podcast. Talk to a professional. But if you're facing real struggles with health, stress, or mental well-being, don't self-diagnose. Talk to someone qualified who understands your specific situation.

And just to be clear: I'm not promoting or discouraging any medical treatment. If you're taking prescribed medication, stick with your plan and trust your provider. This section isn't about replacing that—it's about perspective, curiosity, and showing up more intentionally in a tough, demanding industry.

That said, if we want to talk about fun, fulfillment, and sustainable motivation, we need to understand at least a little about how our brains and bodies work—especially the signals they send us and the systems we use to interpret them.

Within our industry, we have often been taught to avoid anything resembling feelings and emotions. We are supposed to be logical, task-focused, output-driven robots and "keep emotions out of it" (see the glossary for a helpful revised, clarified meaning of this overused directive). Our equipment, computers, and robots operate like this and are built to provide exact outputs based on input commands. But we aren't mechanical machines, and (un)fortunately, this isn't how humans work. It isn't how *we* work.

We are much more complex and operate by taking in information from our environment, filtering it, then creating our own mixture of feelings and emotions which ultimately create actions. They are what drive us and give us motivation, energy, and direction. If we didn't have these, ironically, we would never show up for work in the first place. In fact, we would probably never change out of our bloomers and park it on the couch with a bag of pork rinds all day every day.

Our leak, our fun problem, is a human element problem, not an equipment or process problem.

Happiness, fun, enjoyment, and fulfillment are all feelings. So are trust, respect, confidence, and pride. Feelings and emotions, simply put, are the language our brains and bodies use internally to communicate and inherently, intentionally or not, become a part of how we communicate and interact with others. They're generally combinations of neurochemicals and hormones being mixed with external inputs. Happiness, for example, is the word we use for the feeling of four different neurochemicals being activated: dopamine, endorphins, oxytocin, and serotonin.[24]

Our control lies within our mindsets and energy levels, which all incoming and outgoing information gets processed through. These two critical, key components determine how our brains and bodies take in, absorb, and process information from our environment. This, in turn, informs our brains and bodies what concoctions to mix in order to interact best with that environment.

So our leak, and therefore our fix—and our key opportunity to making construction fun again—narrows down to one of these two major pipeline components: the mindset lens filters and the energy gauge.

Mindset Lens Filter Detail

No matter how much external information is flooding into our pipeline, it all gets filtered through our mindset first. But instead of filtering material like a #200 sieve that lets fine silt and dust particles pass through while holding back sand and larger aggregate, I want you to think of our mindset more like a *lens filter*—not just separating by size, but actively sorting, labeling, assigning meaning, and determining value.

---

24 Loretta Graziano Breuning, *Habits of a Happy Brain: Retrain Your Brain to Boost Your Serotonin, Dopamine, Oxytocin, & Endorphin Levels* (Avon: Adams Media, 2016), 13, https://workmindfulness.com/wp-content/uploads/2017/11/Loretta-Graziano-Breuning-Habits-of-a-Happy-Brain-Adams-Media-2016.pdf.

Our mindset doesn't just block or allow material through; it shapes what we see, how we interpret it, and even what we notice in the first place. In some cases, it determines what makes it into our awareness at all.

Which means our brain—our mindset—is telling our eyes what to see, instead of our eyes telling our brain what they're seeing. That shouldn't be surprising when you consider this: we have ten times more neurons traveling from the brain to the eyes than from the eyes to the brain.[25]

If the filtered content is deemed valuable, it will be sent through to our energy gauge to be measured and further evaluated. Content or inputs deemed garbage will be sent through the bypass valve and spilled back out into the environment. That old saying of "in one ear and out the other" would be a good example of the information flow going through these bypass valves.

There are many different mindset lenses we can choose to filter external content through. Each lens is uniquely crafted over our entire lifetimes from our experiences, life lessons, failures, successes, ambitions, and desires. These can be switched out, chosen, and used in the present, but also can be engaged when looking at and thinking about our past and our future.

These mindset filters are located both at the inlet and outlet of these pipelines. The inlet filter greatly influences the admixture of those neurotransmitters and hormones, ultimately creating feelings, emotions, decisions, and motivation. The outlet filter greatly influences our reactions, actions, interactions, and perceptions of others and of the world around us.

---

25 Breuning, *Habits of a Healthy Brain*, 81.

Take stress, for example. Did you know stress is actually good for you? I didn't. That's how I ended up in the emergency room in the middle of the night mid-career. What I didn't appreciate at the time is that stress is simply our body's language trying to communicate that something we care about is at stake, and motivating us to do something about it.

Stress is a *feeling*—a chemical cocktail of adrenaline, noradrenaline, and cortisol. And, as humans, we *need* a certain amount of it. In fact, our ability to feel happy requires a certain level of stress. Endorphins—a key ingredient to happiness—are released once you push past your capacity to the point of distress.

In healthy doses, stress is not only natural, but necessary. It's healthy to fluctuate around the middle—say, between three and eight out of ten on the stress scale, since this is the zone that keeps us motivated, alert, and engaged. It may even be okay to hit levels nine or ten occasionally, and especially during crunch time. But we don't want to stay at these high levels for too long. This is when stress becomes chronic and unhealthy, and we start paying the price.

Dr. Kelly McGonigal, a health psychologist and lecturer at Stanford University, has explored the power of mindsets in her book *The Upside to Stress*. In one of her studies, participants watched one of two three-minute videos: one video described stress as harmful, the other as helpful. Afterward, the participants were put through a mock interview, which is typically a high-stress situation. Everyone's cortisol went up, as expected. But those participants who had watched the stress-is-helpful video produced higher levels of dehydroepiandrosterone (DHEA), a hormone that supports resilience and growth under stress. DHEA also works *with* cortisol to help protect the heart. These results showed that the mindset of these participants had shifted their *biological* stress response.[26]

I learned this the hard way: mindset filters aren't just mental; they're physical, too. The lens we look through when we deal with stress can literally change how our body reacts. If we see pressure as a threat, our system locks up. We get rigid, reactive, closed off. But if we see it as a challenge—or better yet, an opportunity—we stay more focused, steady, and in control. That's really what people mean when they talk about a "growth mindset." It's the opposite of a fixed mindset, which stays stuck and refuses to adapt, even when the facts change. A growth mindset stays open, willing to adjust and grow through pressure instead of just surviving it. And here's the thing: mindsets aren't permanent. They're tools. You can build them, adjust them, toss them, or trade them out when they stop working. The mindset you bring to a situation shapes what you notice, how you respond, and how you feel—on the job and in the rest of your life.

Energy Gauge Detail

As humans, we require energy to operate. Since energy cannot be created or destroyed but only transferred or transformed, and we don't have an endless supply, we must manage our energy closely. This is why the energy gauge is a standard built-in function of every human.

This internal gauge constantly measures and manages how much energy we use, what kind we tap into, and whether we save, gift, invest, spend, or store it for ourselves, as well as when, where, and how to get more. Most of us think of

---

26 Kelly McGonigal, *The Upside of Stress: Why Stress Is Good for You, and How to Get Good at It* (New York: Avery, 2015), 9, https://www.amazon.com/Upside-Stress-Why-Good-You-ebook/dp/B00OI5PGWU/.

energy in physical terms: what fuels us throughout the day and gets replenished when we eat, hydrate, or sleep. While those things support physical energy, they also directly fuel the electrical and chemical energy that powers our brain. This is the energy we rely on to intake, process, and respond to information—our mental energy, or brainpower. The brain uses about twenty percent of our body's total energy, even though it makes up only about two percent of our body weight.[27] That energy demand depends not just on mental workload, but also on how well-fueled and hydrated we are: what we've eaten, how much we've slept, whether our morning coffee has kicked in, and other physiological factors. And unlike fuel levels or crew hours, this kind of energy is much harder to track—especially in construction, where nearly everything we measure is physical. Still, that doesn't make it any less real. In fact, it's being burned constantly, whether we're aware of it or not.

Any information that isn't discarded by our mindset lens—anything that shows potential value—is passed through the energy gauge for further evaluation. This internal check asks whether it's worth spending energy to keep it. Some inputs give us energy, some require energy to process, and some are just junk. If something is flagged as worthless, risky, or too energy-expensive to bother with, it gets dumped, spilled back out through the bypass valve and into the environment. But if the signal has enough potential to extract energy from, or is worth investing energy into, it moves forward into the pipeline for deeper processing.

These energy gauges, like our mindset filters, are also located both at the inlet and outlet of these pipelines and

---

27 John Medina, Brain Rules: *12 Principles for Surviving and Thriving at Work, Home, and School*, 2nd ed. (Edmonds: Pear Press, 2014), 28.

determine how much "oomph" or energy should be allocated to the incoming or outgoing contents of the pipeline. For instance, how much focus or energy are you going to expend in connecting and communicating with the guy seated next to you at the meeting who is glued to his phone? Do you think your brain and body would find that energy expense a worthy investment in creating fun or fulfillment? Maybe! Alternatively, it may decide to forgo that investment based on how much sleep you got or if you ate breakfast and drank your morning coffee.

Mindset and Energy Management: How You Show Up Matters

The mindset lens we look through and the energy gauge we run on determine how we show up. This is true on site, at home, everywhere. You wouldn't walk onto a job without your boots and gloves. Same thing with your chosen mindset and energy level. They're part of your gear. And just like your tools, they require regular maintenance.

Making construction fun again isn't as much about what you say. It's more about how you show up.

Think of it like grabbing your tools in the morning. What mindset have you selected and installed for the day? The mindset you choose will help you figure out how to handle different jobs, different people, and different situations.

Take our work boot example, above. Nobody had to say a word, but we all got the message. Loud and clear. That's the power of mindset in action. You've got a choice in how you see things on any given day—whether you make the jobsite a miserable grind or find a way to enjoy yourself, even when it's tough. And the attitude you bring doesn't just affect you. It spreads—to the crew in the trailer, to the rest of the team, to the project itself. Every time you walk on site, you set the tone.

Here's the kicker: fun, enjoyment, fulfillment ... they don't show up like a catered lunch. You have to build them. You decide whether or not they're part of your day. That decision starts with how you choose to see things and how you react to what's thrown your way. If you're only looking for what's broken, stressful, or out-of-whack, that's all you'll see, but if you're willing to look for the good—even when it's buried under paperwork, rain delays, and busted equipment—you'll find that, too.

When you ignore these filters, they get caked with dust and clouded. Pretty soon you can't see anything further than the end of your nose. It's frustrating. In this business—where things move fast and problems come faster—you've got to clean these filters. Sometimes you need to rip them out and replace them. Sure, it's easier said than done. Nobody loves chiseling out an old filter that's rusted in place, but with a little grit and elbow grease, you can do it. You just have to be willing.

Some old filters are valuable and might still work, but some are useless. Worse, they might even be holding you back. That's why you've got to stay sharp, stay open, and be ready to swap them out. It's no different than keeping your tools in good shape or staying current on the latest codes.

Mindset Filter Maintenance

I haven't always seen construction as fun, enjoyable, or fulfilling. There have been many times I've been miserable and felt defeated by and fed up with this industry. A very notable time was during the Great Recession. My employer, like many other construction firms, had to lay off a handful of PEs, PMs, and supers to survive and endure as a company.

I had just started that multi-year transition from PE to PM, but because of the economy I had to grow up and assume the responsibilities and role of a PM quicker than anticipated. I had to take over management of two schools undergoing major renovations and additions spanning over two summers, with a reduction in work during the school year while the buildings were occupied. I took over these projects about a month before the second summer started, and it felt like switching drivers at highway speeds. Despite all the other typical challenges associated with this type of project, within a week of me taking the wheel, our electrical subcontractor informed us that, unfortunately, he had to close his doors and file for bankruptcy and could not complete the contracted work.

Fast-forwarding through the multitude of dramas, challenges, change orders, back charges, close-call safety incidents, and arguments, these projects were indeed finished in time to occupy in mid-August. Success, right? Not really. Far from making any sort of profit on this project, we paid the client for these remodels. While these projects were financially scarring to our team, the relationship scarring was worse. I had some of the most challenging ethical dilemmas and difficult conversations I have ever had in my entire career. We all wound up grappling over the minuscule sums of money left in the project. Team members were frustrated and fed up with everyone else on the project, and it was a struggle to drag ourselves past the finish line.

I dreaded driving to work every day of those miserable years. It was not only because of the dreaded interactions with my own team members, but also the nagging expectation that any day I was going to get fired despite me putting 110 percent into my work. There were many times I considered whether I should save everyone the grief and energy and just quit. But I stuck it out, and I was allowed to keep my job. This isn't a rosy lens to look at the industry through. And in situations and times like these it can be challenging to remember that I have the power and control (and responsibility) to change my view—to install a wider lens, to remind myself of the bigger picture or larger overall goal every so often.

Doing this was necessary, not just for my own experience and outlook, but for the entire team. This was a dark time in my career. As I've mentioned before, it was during this period that I fell into depression and burnout. I was anxious and unpleasant to be around. I was frequently on site, jumping in to help extinguish all the small fires and problems, thinking I was helping. But in reality, I was just fueling those fires with my cranky and spicy disposition.

To be clear, having a project manager or project engineer on site can be incredibly helpful. Many superintendents I've spoken with over the years say they'd love to see their PMs or PEs more often. But the key is showing up as a partner, not as an uninvited second superintendent. It's about supporting the field, not working against the person leading it. I wasn't aligned with the site leader. I was hovering nearby, unintentionally making things harder.

Looking back, I was deep in survival mode. If I had swapped out the lens instead of trying to grind through with the fogged-up, rusted one I was using, the entire experience could've turned out differently—for me and for the people counting on me.

Many years later, when I learned about the importance of how you show up, and the power and importance of these mindsets, I spent the time chiseling out the old rusty lenses associated with these projects. I now see that these demanding projects were a major growth opportunity for myself. My current lenses help me appreciate the lessons and experiences they have given me.

Some of the challenges and struggles gave core team members the opportunity to come together and work through them as a unit. We chose to collaborate instead of fight, which helped build real trust, mutual respect, and lasting relationships. Unfortunately, some relationships were too damaged to fully repair. But the ones we were able to salvage have held strong—and have lasted over a decade.

Factory Settings

Clouded and rusted mindset lenses can make it difficult to identify and capture benefits and opportunities from our experiences. Many can be mislabeled as threats, which throw us into survival mode, which then don't allow the proper mixture needed for fun. Garbage in, garbage out. But if you're not intentional, that garbage doesn't leave. Instead, it lingers in your environment and, through a limited, cloudy filter, often gets unintentionally sucked right back in your pipeline, keeping you stuck in the same toxic loop.

Our industry is fast-paced, high-stress, high-risk, and high-pressure. Will we choose to see this as good or bad? Do you view these factors as barriers, burdens, or threats, or do you see them as opportunities, gifts, and benefits?

Your behaviors and outputs into the environment can change the ecosystem for better or worse—and, most noticeably, for those working in close proximity to you.

If you don't see construction as fun right now, it is because you are choosing not to. It may not be intentional, and you may be stuck in some sort of threat feedback loop defaulting to survival mode.

While we have the ability to consciously manage and control our mindset filters and energy gauge, there is a caveat. Our primal, survival-focused subconscious has seniority and authority to override our management of these and default to our human "factory settings" at any time. This typically happens when we have corroded mindset lenses or when our energy gauge senses that our energy levels are too low or have been mismanaged and are not investing our energy wisely enough. While this survival mode is essential and extremely helpful in truly threatening situations, it can unintentionally work against us if we get lazy in managing how our mindsets are filtering and labeling the information flooding into our pipeline.

Think about a time when your boss or supervisor chewed you out for doing something unsafe. You knew it wasn't the smartest move, but you figured it'd be fine just this once. Maybe you justified it to yourself: "Come on, I was just grabbing a tool real quick. I didn't put my hard hat back on because I didn't think it was a big deal. And besides, that other guy's standing on the top rung of the ladder without his gear, and nobody's saying a word to him. So why am *I* the one getting yelled at?"

That's your survival mode kicking in—getting defensive, feeling attacked. It's natural. But here's the thing: your supervisor isn't out to get you. Seven times out of ten, they're coming down on you because they *do* care. They want you to make it home safe at the end of the day to see your family and not become another story none of us want to tell.

When we're running low on energy or not paying attention to our mindset, it's easy to miss this. Our brain flips into an us-versus-them mode, and we see it as an attack, when really it's someone looking out for us the only way they know how in that moment. If we don't stay sharp and manage our energy levels and mindsets, that survival instinct can end up working against us, causing us to react in ways that make the situation worse instead of seeing it for what it is: someone having your back.

Rusty, worn-out, or blocked mindsets may be limiting what we can process from our environment, which can frequently and unintentionally cause us to mislabel content and inputs as threats, missing their potential benefits or value. When we choose to see the majority of our incoming content as threats, it throws an alarm to switch to survival mode.

When in survival-mode, our brain initiates a type of "lock-out-tag-out" of our fun-seeking opportunities and executive functions: critical thinking, impulse control, problem-solving, decision-making, emotional regulation, etc. It essentially puts a spending freeze on our energy stores to ensure they are reserved and used for survival-only reactions—freeze, fight, flight, fawn—until it can convert or find and fill up on more energy.

1 **Protect Energy Stores: Fight or Flight**

Run from or avoid situations, environments, or people (flight), or fight physically or verbally to protect your energy (fight).

2 **Conserve Energy: Freeze or Fawn**
Stop what you are doing, rest, or sleep (freeze), or just go along with whatever everyone else is doing or saying, right or wrong, whether you agree with it or not (fawn).

Think about a time when you're in the zone on an important task, making great progress, and then someone interrupts you with a question. It's like the momentum you finally built comes to a sudden stop. You might snap at the person for interrupting you, not because you're rude but because you're instinctively trying to protect your energy efficiency.

Another example is when you start to feel sick, but you can't *afford* to get sick, so you keep going and try to ignore the symptoms or pain. Sometimes this works and we feel better. Other times, our brain takes over and makes it so we can't even get out of bed until we rest and recover.

These are all examples of rusty, blocked, survival lenses focused on squelching supposed threats. And until we get more fuel and are at a comfortable energy storage level, our brain will have our "thrive"-seeking mindsets locked up.

Our factory settings—survival mode and threat detection—are obviously very important to our survival as humans. They provide us the hyper focus and concentrated energy to effectively address true threats. These only become ineffective and counterproductive when we are haphazardly spending energy on false or misidentified threats—when we're seeing everything as a risk—instead of investing that energy to find and harvest opportunities in these same environments.

Appreciating Construction for What It Is, Not What You Think It Should Be

Decades ago, I had a polished image in my mind of the construction industry: a gleaming new building and a confident leader in a crisp suit standing in front of it, beaming with pride. "I built this," they would say. "You're welcome."

Back in 2003, this was my mental Polaroid of what it would be like to work in the industry. It's what sold me on saying "yes" to my lifelong career in construction. I believed that if I worked hard enough I would master the industry and in return be rewarded, recognized, and compensated accordingly.

Construction, however, isn't something to be mastered, conquered, or controlled. The longer I've been in it, the more I've realized and appreciated its complexity and unpredictability. It's more like raising an unkempt, ornery, sugar-crazed child. You can have everything mapped out, planned out, and in order—including exactly what demeanor your kid will have, the type of parent you'll be, their future accomplishments, and who they'll grow up to become. Of course, the moment you put the finishing touches on that plan, you may as well crumple it up, blow your nose with it, and toss it in the campfire—because it's not going to turn out that way.

Construction is much more than a job. It is much more than a process. It's more than schedules, budgets, change orders, or the proper way to pour concrete, erect steel, or finish drywall. Reading books about construction won't make you an expert. Working in it for twenty, thirty, or forty-plus years won't make you an expert.

Construction is much more than a job. It is much more than a process. Construction is a lifestyle.

Construction is a lifestyle. It will continually test you, knock you down, and sometimes even knock you out. Yet you remain dedicated to it and continue to get up, stand up, and still put immense pride and passion into it. At times, you embrace this. Other times, you resist it. It's a pain in the neck at times, and a gift and a miracle at other times. Construction is something that you can't help but love, even though some days it feels like it doesn't love you back.

I don't blame anyone for having a glorified vision of what it would be like to work in the industry (or be a parent, for that matter). All we have are romanticized postcard pictures of what we thought construction was supposed to be—and a dilapidated map of what others think the pathway to success in our industry looks like. It isn't until we enter it, and start exploring the massive depths and the components hidden in the millions of cracks and crevices of this vast industry, that we truly grasp the complexity of it.

It isn't for the faint of heart or for people who give up easily. While we cannot control or morph construction into a molded version of what we thought it was or what we think it should be, we do still—and always will—have the opportunity to enjoy it for what it is. This exploration and adventure can absolutely be fun if we accept construction for what it is and stop trying to control things we never have and never will have control over.

I encourage you to dig up your own mental Polaroid of what you thought it would be like or what you saw yourself as becoming. It's important to pull this out and dust it off every so often because this is the spark, no matter how small at times, that ignites the torch of curiosity to enter and explore construction. It may have evolved or changed a bit over the years, but it is still there and an important part of who you are. This reminder of what we thought and what encouraged us to take the initial leap into the unknown is important to keep a tight grip on.

I Was the Problem

In many of my experiences shared throughout this book—my overinflated ego, my submittal boxing match, placing blame, pointing fingers, and seemingly competitive co-workers—I was trying to change the industry instead of appreciating what it already is and all that it offers and provides. I was trying to change others instead of appreciating who they are and the value they bring this industry. I was trying to manipulate and control things I had no control over and no business controlling.

The truth is, I can't control the industry. I can't control other people. But I *can* control how I experience it, how I choose to see it, and which mindset lens I choose to look through at any given point.

I *can* choose whether I see the rewards, opportunities, and benefits our industry provides, or I can choose to see them as barriers, challenges, or stressors.

I have control over *this*!

My choice in how I see all this greatly influences how I show up, and this impacts my interactions and how others experience our industry.

You can't reap the rewards of something you refuse to acknowledge or accept. You can't benefit from an opportunity if you view it as a burden. Stop. Take a step back. Recognize the gifts this industry offers. They're there—but only if you're willing to see them.

Moving Forward

Making construction fun again is much less about what you say and much more about how you show up.

What should be clear by now is that construction is a team sport that requires all of us to participate. Getting out of survival mode brings you back to zero, simply balancing the scale. To truly thrive and tip the scale further away from survival mode, you need to manage your mindset filters and energy gauge. These allow you to recognize, harvest, and reignite the fun, fulfillment, and growth opportunities our industry offers us daily. When we get going too fast or are hyper-focused on the urgent in lieu of what is important, these are easily missed.

Our power in this industry lies in recognizing that fun, fulfillment, and growth are not handed to us—they come from our intentional actions and perceptions. Our industry is challenging, but that is not the real issue: it's how we perceive those challenges and how we interact with our environment and the people around us. Our investment and focus on the human element are what give us the opportunity to work through the challenges in the best way possible.

We understand the problem: that if we want to succeed in this industry beyond the projects we build, then we need to change our mindsets and reclaim responsibility and ownership in how we show up. And in doing so, we must also reconnect with the deeper reason we're here—reminding ourselves of the *why* behind construction. Remembering what we're fighting for and why we keep showing up to overcome these challenges together: because the world can't move forward unless we do what we do first.

STEP FOUR

FIXING THE PROBLEM

6

Making Construction Fun Again

"If you are distressed by anything external, the pain is not due to the thing itself, but to your estimate of it; and this you have the power to revoke at any moment."

MARCUS AURELIUS

LET'S BE REAL: finding fun and fulfillment in construction can feel about as easy as finding a level spot on a muddy jobsite. We're often told to tough it out, grind through the deadlines, and power past the pain. That's the job, right? Well, yes... and no.

As we've explored in the past few chapters, the challenges are real, but so are the rewards. And the surprising truth is that the fun is often found smack-dab in the middle of all that hard stuff: the complexity, the curveballs, the chaos. Making construction fun again isn't about adding ping-pong tables to the break room or throwing a pizza party for the guys and gals on site. It's about how you show up—every single day.

We've discovered how managing our mindsets and choosing where we spend our energy helps us reclaim the best parts of this job. The fun, the pride, the satisfaction—they're all right in front of us, if we're willing to see them for what they are.

This chapter marks the pivot point: from understanding to action.

How Do You Make Construction Fun Again?

You choose to decide it is fun! Sounds easy enough, right? Just wake up tomorrow, plaster a forced smile on your face, put a little extra skip in your step as you walk onto the jobsite, and decide this is fun!

Alright, so it's not that easy. The reality is that most of us have spent years building habits and mindsets that have set us up to survive, but not necessarily thrive. Some of these habits we chose, and some have been slowly built, year after year. Either way, it's time to take a hard look at what's working and what's not. Then we can scrap (or chisel out) whatever isn't working and lay a new foundation that actually works in our favor.

And, no, it won't happen overnight. It's a bit like roadwork: tearing out and repaving sections to reroute them to get where we need to go. But since when has hard work ever intimidated us? Anything worth building takes time—and it's worth doing right. We, of all people, know this. This process is no different. We get there the same way we always do: one step at a time, brick-by-brick, mile-by-mile.

When we finally get there, we will step back and be

proud—not just of our projects, but of what we've built in ourselves and within our industry. We'll find that the work fuels us again. That sense of pride, fulfillment, and fun? It's still there. We just have to be willing to roll up our sleeves and go after it.

When has something difficult ever stopped us? We are doers; we're wired for challenge. We solve problems and challenge ourselves and our teams to build smarter, faster, bigger, and better all the time. If we aren't challenged, we become easily bored. We despise monotony.

There is a term that was coined for this: rust-out.

Where burnout comes from pushing too hard for too long, rust-out is when we aren't challenged, we feel stagnant, or we can't see the positive impacts we are contributing to our communities.

Keep in mind: there will always be parts of the job—and life—that just aren't fun, but are nevertheless necessary. Some tasks may be more boring, frustrating, or tedious than others, and we can't just hand them off to someone else because they are. The key isn't to avoid the hard stuff, but to focus on balance and integration, and recognize when the scale tips too far in one direction or the other.

Every single day you have the opportunity and choice in how you show up. Bringing the fun back to construction takes intention. You have to want it. You have to care enough to make it a priority. You have to want to stay in this industry—because if you don't, you won't. But this choice, this decision, is entirely and completely within your control.

Viktor E. Frankl, in his powerful book *Man's Search for Meaning*, talks about something that hits on the importance of this—that no matter our circumstances or environment, there is and always will be one thing that can never be taken from us: our power to choose how we respond. Frankl learned this the hard way, surviving through unthinkable suffering in Nazi death camps from 1942 to 1945. His message carries weight far beyond his experience, though, offering something we can all take with us to work and life:

> Forces beyond your control can take away everything you possess except one thing, your freedom to choose how you will respond to the situation. You cannot control what happens to you in life, but you can always control what you will feel and do about what happens to you.[28]

While clearly what we deal with on our projects and on the jobsite doesn't begin to compare to what Frankl endured, the lesson he shares is universal. We don't control other people. We don't control when a project hits a snag, when a piece of equipment breaks down, or when the weather throws us a curveball. What we *do* have control over is the mindset we bring to work each day. We get to choose whether we see burdens or opportunities. And, further, how we interact and react based on these choices.

Finding the Balance

As I've said before, working in construction—like anything else in life—is about balance. More importantly, it's about recognizing when we're getting too far *off*-balance. In life, there are always tensions pulling us in different directions—working hard, but not burning out; staying focused, but not losing the joy in what we do. If we drift too far one way, we risk burning out. Too far the other, and we risk rusting out. Neither

---

28 Viktor E. Frankl, *Man's Search for Meaning* (Vienna: Verlag für Jugend und Volk, 1946; Boston: Beacon Press, 2006), 86.

is healthy or sustainable. Finding that middle ground, where we're pushing ourselves, moving between these tensions—that's where the work becomes fun, rewarding, and fulfilling.

But also recognize that trying to achieve perfect balance isn't realistic—or enjoyable. Walking a tight-rope isn't the goal here. We can't experience the full enjoyment of happiness without knowing what sadness feels like. Or the full benefits of accomplishment without experiencing failure. Life wouldn't have the amazing wins and ups without the losses and lows.

Hard problems can be both fun and fatiguing, but strong teams with the resources they need can tackle problems without being crushed by them. We are constantly being asked to be more efficient, but often that doesn't give us the time we need to build trust and camaraderie within our teams, or give us the opportunities to fail, learn, and fix mistakes.

We often look back on the good ol' days of construction being simpler and more collaborative. But what we really miss is the camaraderie that helped people handle stress, not an era free of stress.

We're not going to flip a switch and fix the whole industry overnight, and we're not going to fix it alone, either. We *do*, however, get to choose how we show up and contribute. That choice? It makes a bigger difference than most of us realize.

We can choose to build trust. We can choose to have one another's backs. We can learn to spot when someone's getting buried by stress—whether it's ourselves or our teammates—and then step in before it turns into full-blown burnout.

The next sections are packed with tools you can use immediately to make construction fun again. But this isn't about walking onto the jobsite tomorrow pretending like you've got it all figured out. Nobody does. What you do have is a powerful tool to help you get there: your mindset.

Some days, you'll know exactly how to use it. Other days, you're still figuring it out, and that's fine. Every day you get another shot to tighten things up, to show up better, and to get better results—for you, your crew, and our entire industry. Every email, every site meeting, every interaction is an opportunity to move things in the right direction, and even if it's just by a little bit, it all adds up.

There are five main steps to start contributing to making construction fun again that we'll cover in this chapter.

1. **Stop it:** Quit mislabeling the problem or blaming others, and start focusing on what you can control.
2. **Navigate it:** Navigate through or past the toxic contents of the ecosystem, minimizing the negative impact while maximizing benefits.
3. **Find it:** Identify what makes work enjoyable for you and seek it out.
4. **Create it:** Build moments of fun, fulfillment, and camaraderie into your day.
5. **Activate, defend, and maintain it:** Be intentional about keeping the spark alive in your career and use communication to pave the way.

Construction isn't perfect. It never will be, but in its swampy challenges and chaos lies a unique fulfillment. The fun is there if you look for it—but you have to be looking for it to find it.

Let's dig in.

1. Stop It

Stop waiting for someone else to fix it. This industry isn't a restaurant where you can flag down a server to send back your plate because you didn't get what you ordered. Construction doesn't come with a secret team of professionals ready to swoop in and make your job easier, your co-workers more cooperative, or your clients less demanding. If you're looking for someone else to make it better, I've got news for you: you're it. You are the change.

Stop waiting for the "perfect time" to take action. There isn't one. If you're holding out for a break in the schedule, a quieter week, or a magical moment when everyone's in a good mood and open to change, you're going to be standing there forever. Progress doesn't tap you on the shoulder and politely ask if now's a good time. You have to step into it, exactly as things are—chaotic, messy, behind schedule—and start where you are with what you've got. No one's going to build this for you.

Stop expending energy on things you have no control over, and start investing it in things you do. We waste so much energy fighting battles we can't win: the weather, supply chain delays, or decisions made in a conference room miles away from the jobsite. It's exhausting—and it doesn't change a thing. What you *can* change is how you show up. Your attitude. Your behavior. Your decisions. You control how you interact with your team, how you lead, and how you respond when things go sideways (and there are times they *will* go sideways). That's where your influence is. That's where your energy belongs.

Stop thinking you can do it all yourself. Stop trying to do it all yourself. As I've said throughout this book, construction is a team sport. You weren't meant to shoulder the entire load, and you'll burn yourself out if you try. Some of us were incorrectly taught somewhere along the way that the best leaders are the ones who carry everything on their backs, never ask for help, and micromanage every detail. That's not leadership. That's isolation. It cuts you off from your greatest resource: your people. Engage your teammates. Trust their abilities. Patch and repair the human element pipelines that connect you. Lean into their strengths, even when they're different from your own—*especially* when they're different from your own.

Stop pointing fingers and blaming circumstances—or other people—for what's not working. Sure, it's tempting. The incomplete drawings from the design team, the budget overruns, the subcontractor who didn't show up. But blaming others only creates more division and slows progress. Instead, own your part. Take responsibility for how you show up, how you communicate, and how you lead. Be the example you want others to follow. Create the kind of environment you'd want to work in, even if no one else is doing it yet. *Create it especially if no one else is doing it yet.* Behaviors are contagious.

Stop ignoring the small stuff because it feels like it doesn't matter. It matters. A joke at someone's expense, a crew member who's always the last to know what's going on, a frayed relationship that's "not worth fixing"—those things build up.

They become the cracks in the foundation that eventually collapse the structure. Address the small things. Have the hard conversations early so you can start addressing challenges early. Show up for your people when it's inconvenient. It's in those moments that trust is built.

Stop feeding the cynicism. We've all stood in the trailer or around the conference table and listened to the griping. "Management doesn't care." "The client's impossible." "This job's a dumpster fire." Heck, sometimes we might feel the same way and agree! And sure, it can feel like camaraderie to join in, but it's not. It's a weight that drags the whole team down, whether you realize it or not. You don't have to be a cheerleader, but be a part of the solution, not the problem. It can be helpful to acknowledge the challenges and their struggles, but then insert light or opportunity into the discussion. "I agree. We have had some challenges, for sure! But we have some fantastic, smart people on this team. We've overcome challenges like this before, and we will do it again." This can help influence mindsets and build the motivation and momentum to come together rather than tear each other down. This also impacts how the team goes home at night or, after the project is completed, how the team will look back at it: with either pride or resentment.

Stop dismissing your own influence. You don't need a title to lead. You don't need to be the most experienced person on site to set the tone. Whether you're an intern, a laborer, or a senior superintendent, how you act, react, and interact influences everyone around you. Don't underestimate that. If you want a better team dynamic, be the one who models it. If you want more respect, give it first. If you want better communication, start by clearly communicating yourself. You have more power than you think—you just have to use it.

Stop pretending you're not part of the problem. This one stings a little, I know. But none of us get through this life, or this industry, without making mistakes. You might not have created the mess, but you're here in the middle of it, and you're either helping to clean it up or making it worse. Take an honest look at your role. Are you showing up late? Cutting corners? Dismissing someone's concerns because it's inconvenient to listen? We all do it sometimes. The key is recognizing it and changing course. Take ownership of your actions and their consequences—both the good and the bad. Be part of the solution.

Stop thinking it's supposed to be easy. It's not supposed to be easy. It's supposed to be worth it. Construction is hard. Leadership is hard. Building things that last—whether it's structures or teams or trust—is hard. If you're in this industry, you already know how to do hard things. You've already got the grit. Now it's about using it on the right things. On the people. On the relationships. On the culture. That's where the real work is. That's where the real rewards are.

2. Navigate It

Narcissistic supervisors, Machiavellian colleagues, misused technology, and long hours can most definitely contribute to the toxic bile floating around in our industry's ecosystem. We don't have control over these. We *do* have control over how we navigate these—how we react to and process these if they inadvertently get sucked into our pipeline and sneak past our filters. Our control lies in where our energy and focus go.

Continuing to blame and complain just feeds these problems, giving them more of your focus, attention, and, therefore, more power. There will, of course, be times that we have to navigate through these challenges. We will have to put in long hours, extra effort, or have hard conversations every so often, but don't let it consume or digest you. Remember to step back, take a breath, and remind yourself of what you have control or influence over, and navigate through it instead of getting stuck in it.

While, yes, problems certainly are not fun, they are not what will significantly change the fun factor of our industry. They are symptoms, not causes. Unmanaged, these symptoms can all build on one another and clog up our mindset filter, which reduces the opportunity for any optimism or potential to enter the pipeline. We have to be intentional in cleaning these filters, for making space for the positive aspects, otherwise muck and the trash at the bottom of the swamp will be the only contents flowing through that inlet gate. At that point, we will need to filter through a lot more junk to harvest the small bits of nourishment we can gain. This is our choice, our control, and our responsibility.

Self-Serving Supervisor or Self-Absorbed Co-Worker

Your control lies within how you interact with those around you and your surrounding work environment. You may not be able to change the entire industry all at once, but real progress is made through influence—not force, complaining, or resistance. Change takes time, effort, patience, and persistence. The best way to move a mountain is one stone at a time. Brick-by-brick, person-by-person, we *can* make construction more fun, enjoyable, and fulfilling. Does this mean we should all just put up with selfish supervisors or toxic individuals who can't control their emotions or their temper? No—not at all. That right there is a lack of emotional intelligence. And it's one of the biggest things holding crews and companies back.

Rarely do we get to choose our supervisor, the projects we're assigned to, the people we drive next to on the highway, or the team members we have to work with on site. But what we can control is how we respond, react, and interact—especially by limiting how much those people affect our mindset and our ability to enjoy the work.

We also need to recognize that most people's main goal in life isn't to be a tyrant leader or obnoxious co-worker (okay, for some of them it is), but the *majority* aren't waking up in the morning with that as their life's mission. Neither are they going home at night, smiling and patting themselves on the back for surpassing their record of pissing people off that day.

While we don't need to embrace this negative behavior or toxic leadership style, we *do* have a couple of options:

1. **Interact more with people who energize you.** Their positive interactions can outweigh the negative ones.
2. **Minimize interactions with toxic individuals,** energy leeches, and those who drain us.
3. **Try to understand their perspective.** Some of the toughest supervisors become valuable mentors once you break through their rough exterior.

If you are struggling with a dictator-like supervisor or client, changing your mindset may not magically change them and make them more enjoyable to work with. But it *will* influence how you interact with them. This applies not just to difficult individuals, but also to challenging situations. Your response determines whether an experience is tolerable, enjoyable, or unbearable, and your interactions are the key to navigating these experiences, people, and situations.

Technology

Technology is a tool, not a replacement for human interaction. If we start relying on it too heavily or start mistaking it as an equal substitute for in-person interactions with our peers, we get into trouble and our brains throw temper tantrums.

Think of technology like coffee or energy drinks. They can be great in the short term, but you can't survive on them alone. At some point, real nourishment is needed as fuel. The same kind of problem happens when we rely too heavily on our tablets, project management software, and laptops, forgoing in-person (or over-the-phone) interactions. Technology is fun in the short term, but it doesn't provide us with the more nourishing, longer-lasting, human-element fun, fulfillment, and enjoyment that in-person or over-the-phone interactions provide. It's just how we are built and how we operate as human beings.

Approach technology as a tool. Use it if it makes your job or the management processes easier. Just like advancements in tools, there are some extremely helpful and valuable technologies that can make our lives and work easier, therefore allowing us to spend more energy on building and maintaining the human-element component.

Don't, however, overlook the ways technology can actually bring more fun and satisfaction to the job, too. For example, some project management software can automate tedious paperwork or repetitive scheduling tasks, preventing you from rust-out, freeing you to invest more energy on the human and fun aspects of planning and building. As long as we use technology mindfully—balancing digital tools with in-person communication—they can reduce drudgery and contribute to a more enjoyable work experience.

That "Other" Generation

We seem to have permanently embedded rifts that are wedged between the different generations, causing an us-versus-them approach to life and work. If we always label the older generation as unreasonable curmudgeons and the younger generation as lazy, entitled tech extensions, then we aren't going to make any measurable progress in strengthening those human connections. It takes all of us to build our world. Until these rifts are removed and we weld our generations back together as a larger industry, we will not and cannot experience the full enjoyment or fulfillment of our work.

Each of these generations learned how to interact with the world based on what the world decided to teach them in the prime brain-development years of their lives. This doesn't make any one generation wrong or right, better or worse. Just different. These differences are what make our industry so good at problem-solving and overcoming the challenges thrown at us. Instead of resisting the younger generation's fresh ideas or the older generation's wisdom, embrace the balance and value they both provide.

Problems can't be solved with one way of thinking. It takes many different ideas, approaches, perspectives, and experience all combined, considered, and crammed into one large ball to throw at the problems we encounter on our projects. If we all acted and thought the same way, I guarantee you we would be terrible at problem-solving and we would not be working in this industry.

Yes, it can be challenging to learn the new language of the next generation entering our workforce—or for newcomers to learn the language, accepted behaviors, and cultural norms of the construction industry. But the value of doing so far outweighs any inconvenience.

An equilibrium will emerge, just as it has with every generation that has come before. It takes time, patience, and a willingness to lend a helping hand to younger workers as they adapt to and navigate our complex ecosystem. At the same time, it takes patience, respect, and understanding for the oldest generation as they transition into retirement. It takes *all* of us.

50 Cents Less per Hour

Money matters. There's no doubt about it. It puts food on the table, keeps a roof over our heads, and makes life less stressful. When it comes to real happiness, though, it's not just about the paycheck. Studies show that once you've got enough to cover the basics and feel secure, having *more* money doesn't make you a whole lot happier.[29] What really makes life good is what that money lets you do—like taking care of your family and taking pride in a job well done. When you find purpose in your work and life, that's where the real satisfaction comes from.[30]

We all think money will buy us the life we want. Sure, it helps—but if we're not smart about how we earn, save, and invest it, we'll end up stuck chasing overtime or jumping from job to job for a sign-on bonus and a fifty-cent raise. It's

---

29 Daniel Kahneman and Angus Deaton, "High income improves evaluation of life but not emotional well-being," *Proceedings of the National Academy of Sciences* 107, no. 38 (2010): 16489–16493, accessed January 29, 2025 https://doi.org/10.1073/pnas.1011492107.

30 Carol D. Ryff and Corey Lee M. Keyes, "The structure of psychological well-being revisited," *Journal of Personality and Social Psychology* 69, no. 4 (1995): 719–727, accessed March 17, 2025, https://psycnet.apa.org/doi/10.1037//0022-3514.68.1.127

about making your money work for you so you're not always working for it. Interestingly, the feel-good feelings are what our brains and bodies actually crave when we're motivated to make more money:

- The feeling of security (covering food and shelter)
- The feeling of enjoyment (like a family vacation or overlanding adventure)
- The feeling of pride (ownership of a new SUV or truck)
- The feeling of calm (with new fishing gear or a camper)
- The feeling of energy (tickets to a football game)
- The feeling of authority or success (a new title or promotion)

Being motivated by money isn't wrong. In fact, it is perfectly acceptable. Understanding the underlying want or need that is making you chase after money will help to ensure you aren't continuing to run on the treadmill with a hundred-dollar bill dangling in front of you. Yes, you have to earn a certain amount to survive, but at a certain point, what, ultimately, are you trading your energy for? Is it getting you any closer to what you want this to be?

Like any resource, we can exchange a certain amount of energy to earn money to pay for or acquire the resource, but we can also invest energy directly into growing and harvesting the resource ourselves. Therefore, energy is the true investment, with enjoyment being the resource (fuel) acquired. Money is simply one way to purchase or acquire that resource, that fuel.

Long Hours

Some trades, tasks, or projects inherently require longer hours—like pouring concrete in the wee hours of the morning to avoid the relentless heat that would make it set too fast. What we do still largely depends on sweat equity to support and maintain essential elements of our society. Like all the items we touched on above, however, there is a balance.

Believe it or not, there are those in our industry who *enjoy* longer hours. Maybe not all the time, and not excessively, but this is what energizes them. It may be overtime hours to earn extra cash for something they are saving up for. It may be that working longer hours makes them feel valued and helpful, which can be fulfilling. It may be to avoid a toxic home environment. It may be that the type of work they do keeps them out of trouble, avoiding idle hands or an idle mind. We've all got different reasons, preferences, and limits.

Sometimes it's the season of life you're in, or maybe it's just the stage of the project. Regardless, it's on *you* to notice what's happening and either endure it or do something about it. Consider the role, tasks, specialty, or sector of the industry you're in, and ask yourself whether it fits your trajectory, energy level, and skills. If you hate working twelve-hour shifts, don't sign up for a gig that requires them. Some of your teammates love that pace, so let them thrive there—no point in opting in, then grumbling and dragging everyone else down.

Sure, there are nine-to-five jobs out there. You might be stuck in a poorly lit cubicle, clacking away at data entry all day. You might be working from home in your sweatpants all day. But for many of us, that kind of work would feel longer than a twelve-hour day on site. We chose construction for a reason: it's hands-on, it's out in the field, and we get to work with awesome people and build stuff that matters.

That said, we shouldn't be requiring or pulling ten- or twelve-hour days *all the time.* If that's become the standard on every project, day in and day out, that's usually a sign of poor planning or management. Sure, certain projects might demand overtime for a while—again, like being at a high stress level for a limited stretch—but constantly burning the candle at both ends can lead to burnout or injuries. We only have so much energy, so we need time to rest and recharge to keep our heads in the game. If your trade requires this type of work, then consider rotating shifts or creating fatigue-management plans.

Keep in mind that long hours and demanding work aren't the whole story. Your mental health matters, too. If you're feeling overwhelmed or dealing with a rough time at home or in your personal life, talk to someone you trust. Company culture plays a big role, too—some places require and advertise nonstop overtime, some don't *require* it but *expect* it, and others encourage more balanced or flex schedules. Plus, if you're in a union, there are rules and labor laws that can protect you in certain situations. Good communication with your team and supervisors can open the door to more flexible hours or better planning.

Finally, time (energy) management—like rotating shifts or delegating tasks—can prevent burnout. All of this helps you stay safe, sane, and more balanced, which ultimately makes the job more enjoyable.

Work-Life Balance

Work-Life balance is elusive, but that's because it's not possible to completely split your life into two neat halves. There is no way to fully separate work and life. You are the same person whether you are walking onto the jobsite or walking through the front door of your house. Try cutting water in a bowl with a knife. It's just going to flow back together.

I prefer the term "work-life integration," because who you are at work should build upon and complement who you are outside of work—your skills, hobbies, community, network, and experiences. Your home life influences your work life, and vice versa. This doesn't mean using the same language or spreadsheets you use to manage your crew on your spouse or kids. That likely won't go over well.

If you insist on striving for that work-life balance and are trying to keep the two separate but in some kind of equilibrium, I'd encourage you to at least *increase the tolerances.* Think of it more as an eight-foot sidewalk rather than a tightrope. We all veer a little left or right sometimes, putting more energy or focus into work or home. Just pay attention to when you're drifting too close to the curb—too close to the end of one of those tension points. If you do, make an adjustment. That way, you won't fall off and break your ankle. Communicate with your significant other and the important people in your life so they know to expect a little less energy at home for those times when work needs more of your attention. Alternatively, listen to your spouse, family friends, or workmates when they are warning you that you are getting too close to one curb or the other. Sometimes they see it before we do.

FINDING THE RIGHT COMPANY CULTURE

Alicia Ready has spent nearly two decades in commercial general contracting, growing from project engineer to a leader known for her commitment to people and culture. Drawing on her background in project management and her passion for mentorship, she now focuses on developing people and helping teams grow. Below, she shares her insights on evaluating company culture—and how finding the right fit, and surrounding yourself with the right people, can make all the difference in your career.

"There may come a time when you recognize that part of the 'loss of fun' you're feeling is rooted in the environment you're currently in.

"When I was just starting my career, I didn't really know what to look for in a company—at least not from a cultural standpoint. I thought culture was just a buzzword, or maybe something they slapped on the wall in the break room. I didn't realize how much it would matter over time.

"As I grew in my career and in my own awareness, I started to see it clearly: not every company is created equal. And those catchy phrases they print on the website—integrity, trust, relationships—those aren't always backed up by real actions.

"What matters most is how a company *acts*. Every decision they make, how they spend their money and resources, how leadership treats their employees, how co-workers treat each other—that's the real culture. You could call it 'culture currency.'

"Every choice a company makes is either a deposit or a withdrawal from that account. Healthy cultures build up currency with things like transparent communication, mutual respect, and thoughtful decision-making. Toxic cultures make constant withdrawals—poor communication, low accountability, fractured trust.

"And here's the thing: not every culture is going to be right for you. You're not going to change an entire company's culture single-handedly, and it's not your job to try. It's a tall order to expect yourself to undo years of ingrained behaviors and mindsets. Your energy is better spent finding a culture where you can thrive, not fighting against one that doesn't fit you.

"Sometimes, you outgrow a company's culture. That's okay. Sure, it's scary to make a move, but sometimes the grass *is* greener on the other side. Once you realize an environment isn't a good fit, it's better for everyone to have the awareness that it's time to take the initiative and part ways. You want to leave a company on good terms and acknowledge them for the ways they have helped you grow and learn on your journey.

"It's not the company's job to be everything for everyone. Their job is to run their business. Your job is to know yourself well enough to find an environment where you can thrive and leverage your strengths.

"If you're not sure how to tell whether a company's culture is healthy or a good fit, take the time to find out. Interview them as much as they interview you. Ask open-ended questions like:

- Tell me about a time when your company's values were tested. How did you respond?
- What does your mission statement really mean to you in day-to-day work?
- Which of your company values is your favorite, and why?
- How do you see leadership demonstrate those values?
- What does an ideal culture fit look like here?
- What legacy does your organization want to leave for the industry, your employees, and your community?

"Pay close attention to how they answer, not just *what* they say. Do they stumble when you bring up values? Do they deflect when you ask about leadership? Their tone can tell you everything you need to know.

"And don't ignore industry reputation. Subcontractors, owners, and design teams often have an unfiltered view of how a company treats people—inside and out. If you hear consistent negative feedback, take that seriously.

"Culture matters more than we give it credit for. It shapes our energy, our growth, our well-being. And while you can't control a company's culture, you *can* control where you choose to work.

"Know your non-negotiables. Trust your gut. And don't settle.

"The right culture does exist. And life's too short to stay miserable in a toxic one."

3. Find It

Gifts and Fun Packages Provided by Construction

Fun, enjoyment, and fulfillment are feelings that can only be created and concocted by you—using the content you choose to gather from your environment and filtering it through your mindsets and energy gauge.

Just like any other resource, it requires energy and intention to invest in, dig for, find, and harvest these ingredients from your work environment, but in return they provide you with the nourishment of producing fuel and motivation.

By actively engaging and investing in our work environment, we can uncover the hidden gems of fulfillment and fun, ultimately enriching both our professional and personal lives.

These opportunities are everywhere, all around us, and are presented to us every day by our industry. If we are too busy, too stressed, or too focused on the negatives, we will miss them.

First you need to be able to identify them. Determine what is fun and fulfilling for you. Ask yourself questions like:

- What am I looking for?
- What do I like best about what I do? What do I like least?
- What has been my most significant or impactful accomplishment in the past three years (personal, professional, private, or public)?
- What would I ultimately like to be known, remembered, or recognized for within my career?
- Who would I most like to help or provide a positive impact for?
- Am I pointed or taking steps toward the right direction?
- What am I sucking up through that inlet gate of my human element pipeline? Is it the discarded muck at the bottom of the swamp, or is it nourishing and positive content that will contribute to getting me to where I want to go?

Once you have determined what you need to make this industry fun again for yourself—and what inherently satiates us as human beings—then you can be more intentional about what you are looking for, harvesting, and absorbing.

Here are just a few of the human-element nutrients provided by our industry:

1. **Endless opportunities to learn, grow, advance, and explore,** which continue to push and challenge us to make ourselves smarter and stronger.
2. **Opportunities to build robust, lasting relationships** with some amazing individuals.
3. **Hands-on work** with tangible results from our sweat equity.
4. **Advanced problem-solving and mental exercises**, which provide endless learning, as well as personal and professional advancement and fulfillment.
5. **Potential to mentor and impact** more people than many teachers or coaches, simply by doing our jobs well.
6. **Being part of something larger than ourselves**—taking pride in what we do. Accomplishing amazing things and building phenomenal structures. Providing essential life-sustaining systems for our communities which enable society to survive and thrive. This is what all of us, primally and instinctively, are motivated toward as humans. We want to create legacies.

And underneath all these individual opportunities lies something even deeper: purpose. Construction gives us something to fight for—something real and meaningful. It gives us a reason to come together, to endure the hard days, to grow stronger through challenge. We aren't just building projects; we're building the foundation of society. Reminding ourselves and reconnecting with this bigger "why" taps into a source of energy and resilience that helps us overcome the hardest parts of this job. We remember that the work matters, and so do we.

Construction provides an amazingly nourishing career, if we aren't too busy or going so fast that we miss all the opportunities. What we do, what we create and contribute to, is life-sustaining, not only for ourselves but for our world. This is important for all of us to remember: to pick our heads up from our day-to-day tasks from time-to-time, take a breath, and recognize these little nuggets of opportunity.

Not Enough Time: Demystifying Time-Management

When we are busy trudging along, trying to address the problems and emails piling up every hour of every day, it can be hard to find these opportunities. To relieve this, we are supposed to "manage our time better," which apparently is some elusive magical skill that no one actually acquires, yet somehow are expected to implement.

Honestly, I hate the term "time management." It doesn't make sense, and it lacks context. Time is a unit of measurement. We don't have any control over time, and therefore we cannot manage it. Being asked to manage your time better is the equivalent of being asked to manage a cubic yard better.

Instead, think of this as *energy management*. We have control over and can manage the amount of *energy we expend* and *what we focus that energy on* within a time unit (hour, day, month, lifetime).

Yes, we are all hard workers, but we can unintentionally and unknowingly put our energy into the wrong things or even in the entirely wrong direction! Our energy has a massive impact: good, bad, intentional, or not. We can find enjoyment or torture in anything and everything. It's our choice what we decide to take from those experiences and moments.

Simply put, expending our energy isn't the same as investing it.

How we react to and approach—and what we take away from—an experience, task, or moment is one-hundred percent our choice. We control that and, further, we own it. It isn't until we take ownership of the responsibility of this choice that we can fully benefit from it.

4. Create It

We cram so much into our schedules for the sake of efficiency and productivity that it has created an industry-wide intolerance for anything close to resembling waste. We give a standing ovation to anyone who can trim down project timelines, budgets, meeting agendas, and blueprints—even how we organize tools in the job box. Considering how much waste our industry can create, that's not just helpful, it's critical. It's awesome!

On the other hand, we love this "cut-the-waste" mindset so much that we've carried it over to our personal lives—our kids' after-school schedules, the dishwasher, even our vacations. It feels good to pack the maximum amount of experience and enjoyment into the little time we have at minimal cost (whether monetary or energy). The problem is we sometimes confuse *space* with *waste*, cutting out or filling every gap in the name of efficiency.

When Space Becomes "Waste"

In trying to get rid of every bit of waste, we've accidentally started treating normal breathing room and mind-space as if it's useless. Yes, it's important to eliminate actual waste, but we also need to protect and keep space—both physical and mental—in our work.

By stuffing our schedules to the brim, we leave no room to rest, think, or just breathe. If there's no free space in the day, there's no chance for real human connection or even just a quick mental break. It's like trying to cram ten pounds of crap into a five-pound bag.

Many believe they're being more productive by working straight through without breaks, thinking, "If I'm taking a break, I'm not getting anything done." However, our brains need downtime to process all the info we've been taking in. This idea of space also includes elbow room at the jobsite and enough time in the schedule to finish a task (or project) well—plus mental space to think clearly.

Here's a question: which option helps more when you're behind on work yet overwhelmed?

A Taking a break.

B Doubling down and working until it's done.

Many of us engage B, but the truth is that A—taking a break—is usually the better move. When we pause, our brains go into high gear behind the scenes, sorting and organizing everything we've learned. They also clear out their own "junk," which makes us sharper and more ready to get back at it.

If we skip breaks or fill them with social media or other junk food for the mind, our brains don't get the chance to do this essential clean-up. They'll attempt it anyway while we're still working, but neither that or the skilled task we are working on will be done effectively.

Breathing Room

Think about the structures we build. Without expansion joints in concrete or masonry, or space under columns for shims and grout, these structures would crack and fail. By intentionally installing and incorporating "breathing room" for our materials, they are more likely to crack in designated, intentional locations instead of undesirable or unsafe locations.

The same goes for us, as humans. If we aren't intentional about providing mental or physical space for ourselves to take a breather and allow fun to be absorbed, then over time, that space gets stuffed with trash, dirt, and debris. We need to implement intentionally planned space and room in our lives and our schedules to breathe and think. If we aren't intentional about this, we will crack at an inopportune moment or in a space that is unsafe or a higher risk to our health, career, or reputation.

Consider how can you be more intentional about crafting your career—or even just your day-to-day duties—so they include more of what you love. It may be getting out in the field more often, volunteering to speak at high schools about career paths, or mentoring a younger teammate or industry newbie. Maybe it's carving out time to teach, lead, or volunteer within your community in a way that recharges you. As we all know, if we let the paperwork, red tape, and HR complaints pile up without something to balance them out, it gets heavy. Especially if this type of work is not your jam. So be deliberate about investing in the energizing and rewarding components.

If you're in a leadership role—whether it's as a client, designer, project manager, general superintendent, crew lead, or any position where people are looking to you—how can you create or encourage more space for your teams? Can you build a schedule that allows for breathing room, instead of stacking trades on top of each other? Can you create more "air space" in meetings where people can collaborate openly, ask questions, or throw out their ideas without fear? Can you carve out space by reducing unnecessary meetings altogether? Seriously, does that meeting need to happen? Do *all* those people need to be in it?

Beyond that, who can you give more autonomy? Who can you trust with time and space to decide how best to apply their skills and talents? Sometimes the best thing you can do as a leader is get out of the way. Let them shine. Don't smother the flame you worked so hard to ignite in them.

When people are given the room to breathe, think, and problem-solve, they bring their best to the table. When that happens, you don't just get happier, more engaged teams—you get a better product in the end. Issues get worked out in real time on site, instead of turning into paperwork battles later. That means fewer change orders, less back-and-forth, and a smoother process for everyone involved. It means less rework, less waste, and a whole lot less energy spent putting out fires that didn't need to be started in the first place.

Imagine what creating that kind of space could do—not just for you, but for your crews, your teams, your trade partners, and the industry as a whole. You can't afford *not* to create this space. This is how you start making construction fun again. It allows you to be more aware, more engaged, and to notice and harvest the opportunities.

Reactive Versus Proactive Mindsets

If you don't plan your energy, your mood, and your breaks, the day will be planned for you by everyone else. This can leave you drained and spicy and eliminate any opportunity to harvest any fun.

Think about your workday morning routine. Not only the one that starts once you step onto the jobsite or into the office, but the one that starts the moment you wake up. Within your first hour or two, do you check emails or messages, answer phone calls, scroll social media, and watch or read the news?

Most of us who work in construction will consider this a ridiculously obvious question. *Of course* we're doing these, often within the first two hours of rolling out of bed. It's as common as pouring a cup of coffee. Now these aren't inherently bad or wrong, but there is a problem. All of these are *reactive* activities, which means you're letting something else control your mindset before your brain is even fully awake. It's like giving away the driver's seat on your own day.

Remember, we can't always control what happens around us—especially on a jobsite—but we can control how much energy we give to each situation. To do that, though, we need to have enough energy in the tank. Like a project plan, we should be intentional about how we start our day and how we plan to handle our mindset and throttle our energy.

When we are in a reactive state, we become armored and defensive in maintaining our energy. It is much harder to get out of this survival mode, and it winds up taking a lot more energy than simply maintaining a proactive state. Reactive states do and will happen. Practicing and engaging self-control by intentionally setting our mindsets helps reduce the amount of time and energy used for reactive states throughout the day. We maintain control of our energy.

If we show up to work rushed and defensive, then we're already behind the eight ball. We're reacting to everything instead of being calm, clear-headed, and ready to take charge. If we put enough energy up-front—planning how we'll face inevitable problems—the day usually goes a lot smoother. Sure, emergencies pop up, but when we're in a proactive headspace we burn less energy overall than if we're constantly putting out fires in a panic.

Most of us do not intentionally plan out and set our mindset based on what we need or want for the day. This isn't easy to do. Our brains are tired in the morning, so other people laying out a to-do list for us is appealing. Our brains want that quick energy, like donuts, coffee, or energy drinks, and for something to motivate us like an email that was sent out late the night before. These are quick and easy "feel good" reactions. The problem is that at this point in the day, we aren't controlling the mindset, energy, or agenda that will influence the remainder of the day and week.

To improve how and when we use our energy—and our mindset—we can practice three things:

1. **Self-control:** Decide where your energy goes. Stand firm on your need to recharge and take breaks. Install boundaries.

2. **Self-awareness:** Know how tired or energized you are and what builds you up or tears you down. Figure out your strengths, weaknesses, and when it's time to step on the gas or hit the brakes.

3. **Self-management:** Put the first two into practice. Are you showing up in a way that's good for you and your team?

Learning from Architects: Designing Our Days

We're not here to debate project-delivery methods or how squeezed for time design teams often are, but there's a universal truth in this industry: the more time you spend planning, the better things go.

Building on the above, consider the following:

- Start each day—or week—by making a game plan. Resist the urge to dive straight into your emails; they'll kick you into reactive mode. Give yourself an hour (or even thirty minutes) to plan your day first. Instead of opening your inbox, open your calendar or schedule. Line out what your day needs to look like—and how you need to show up.
- Be honest with yourself about what's on your plate, how much energy it will require, and what could go sideways if you don't leave yourself enough space or fuel to handle it.
- Block out some "pit stops" to recharge—whether that's grabbing a bite, stretching your legs, or just catching your breath. Put it on your schedule. Don't just tell yourself you'll get around to it. I know we are very selfless individuals when it comes to our teams, but if we constantly run on empty, we won't be able to show up as the good leaders and co-workers we want and need to be. Emotional intelligence—knowing how you're feeling and *managing* your emotions—goes a long way here. Saving a little energy for yourself isn't selfish; it's the best way to help the team in the long run.

Fuel and Energy Pitstops

We all know some things drain our batteries and some things fill them up. Make time for environments, activities, people, or situations that recharge you so you don't run on empty. Further, intentionally schedule these recharging sessions which ensure your fun, fulfillment, and enjoyment levels are properly fueled and maintained.

Explore and identify some of these different options and the unique "batteries" that recharge you. Make sure you write down these different options, so that even if you're bone-tired, you can still locate them. This will help to engage when an energetic, inspirational, or motivational boost is needed. Below are some suggestions on different ways industry members re-energize, and I've also included a helpful worksheet in Appendix 2.

Increase your oxygen: Getting oxygen into our system and to our brain is vital for optimal operation. One of the best ways to do this is through exercise. This doesn't mean we have to go and get a gym membership and work out forty minutes a day (there's a risk of burning out if you set yourself a big goal like that too soon). Getting oxygen into our system can be done with smaller and simpler steps, and throughout the day, on the jobsite.

Physical activity and exercise increase the efficiency of getting oxygen into our system, helping our brain function and impacting our self-control. Walk up the roof-access ladder and check on the progress of the RTUs (rooftop units); go to the furthest point on the site to see how the landscapers are coming along. Heck, propose walking meetings that keep discussions focused and to the point but also improve conversation and connection with individuals because they allow you to "work things out."

Go outdoors: Camp, hike, and fish in your downtime. Walk around the jobsite or pay a visit to the site if you work in the office. Being outdoors for just five minutes can help decrease stress, improve your mood, enhance your focus, and boost your self-control.[31] Take a phone call while walking, or intentionally plan a short walk in nature. This will help you increase your oxygen intake, too.

Volunteer: Say yes to participating in an industry panel to encourage youth to consider construction as a career. Coach a sports team. Help at Habitat for Humanity, or head up a community service group within your organization or community. While sometimes these can seem like things we don't have time or energy for, recognize that if it is something you are passionate about, the time spent is an investment, with its return as more energy and enjoyment in your work or home life.

Stay hydrated: It seems like we are constantly being told we need to drink more water. We prefer coffee, soda, and beer to water, because they often taste better and immediately impact our energy levels! But in order for our brain to work the way we want it to and need it to, it requires a lot more H2O. Dehydration affects our attention, planning ability, self-control, focus, memory, and even simple math skills. By all means, drink your favorite energy drink, but then make sure you set yourself up for success by chasing it with a glass of water. An easy trick to get you drinking more water is to match anything that's not water with water, one-to-one; follow each cup of coffee with a cup of water, for example. Your energy levels—and your kidneys—will thank you for it!

---

31 Kelly McGonigal, *The Willpower Instinct: How Self-Control Works, Why It Matters, and What You Can Do to Get More of It* (New York: Avery, 2011), 44, https://www.amazon.com/Willpower-Instinct-Self-Con%20Works-Matters-ebook/dp/B005erIrZe/.

Learn something new: Read books, listen to podcasts, take a class, and invest in yourself! Learning comes in many different forms. Just be intentional about what you feed your brain. Enter every day, challenge, conflict, and conversation with curiosity and with the intent to learn something new or grow from that experience. This gets you into the mindset of learning new things and makes challenging situations easier to approach, communicate, and handle. Even aerobic exercise can be a brain-builder, as it helps improve your mind-body connection.

Spend time with those who energize you: Family, inspirational colleagues, groups or clubs, or even by yourself with a good book. Recognize that certain activities or individuals energize you more than others. Conversations or working with certain people may make you feel energized, so schedule coffee with them! What are ways you can schedule intentional conversations or engineer ways to work with these energizing individuals more frequently?

We know that stress happens, so here are some quick ways to recharge yourself:

For a very quick five-minute recharge or de-stress:

- Find a space where you will not be disrupted or interrupted. Silence your phone.
- STOP. Unclench your jaw. Soften your forehead. Relax your shoulders.
- Put in some earphones, or jump into your vehicle, and access one or two of your favorite songs.
- Sing along and breathe deeply. This gets us breathing oxygen into our system and re-engages our brain's

executive functions. Music can create hormonal changes that improve your mood.[32]

For a longer five- to fifteen-minute recharge or de-stress:

- Find a space where you will not be disrupted or interrupted. Silence your phone.
- STOP. Unclench your jaw. Soften your forehead. Relax your shoulders.
- Take five to ten slow, deep breaths. Breathe in through your nose and into your stomach, and then out through your mouth.
- Visualize one of your recharge spaces, places, or people. It might be a lake, or the beach, or the mountains, or at home with the family—anything that helps you feel recharged. Feel it, smell it, hear it. Fabricate the space in your mind to the fullest, and be there for five to fifteen minutes.

If you don't plan your energy, you'll lose it. When you don't map out your own energy or build in breaks, you'll end up blown around by other people's problems and agendas all day. That's a recipe for burnout and zero fun. Sure, there's a chance you may end up where you wanted to be anyway, but more likely you will end up where everyone else has pulled you throughout the day—and end up exhausted, cranky, and looking back at cumulative days and weeks that are far from being considered fun.

32 Medina, *Brain Rules*, 216.

Bottom Line

1. Space isn't waste. We all need breathing room, just like our structures.
2. Breaks aren't lazy. They refuel us so we can get more done.
3. Plan your mindset toward how you need to show up for yourself, your teammates, your family, or your community, so you stay in control, rather than letting the day run you.
4. Recharge in ways that truly give you more energy.

We want to cut real waste in the construction world, but we can't confuse that with the space we *need* to stay healthy, happy, and on top of our game. When we allow for breathing room—both in our structures and in ourselves—we're more likely to succeed and have fun doing it.

5. Activate, Defend, and Maintain It

Now comes one of the most challenging parts: *activating, protecting, and maintaining* this desirable and sustainable state. We can invest a lot of energy discovering the fun aspects, harvesting the enjoyment, and creating and building our own fulfillment roadways—but if we can't activate, defend, or maintain it, then it will quickly become populated with heavy traffic and barriers or overgrown with weeds that you will have to expend unnecessary energy in clearing.

How do we do this? Communication.

An entire book could be written about all the nuances, depths, crevasses, and veins of communication—how to communicate, why to communicate, when to communicate, and whether you're communicating with the right person to get your message across effectively. Heck, my entire business is centered around this step. Until I write that book, we'll need to make do with a brief flyover of how to communicate—so you can activate, protect, or maintain your recently harvested or hard-earned fun, fulfillment, and enjoyment within the industry.

More Than Words

First off, don't think of communication as just words (remember the work boot example from the previous chapter?). Some studies suggest that words account for less than ten percent of how we actually communicate. While the right words can help, what matters even more is nonverbal communication—which makes up the remaining ninety percent or more. It's your tone, body language, and presence that either support your words or speak in place of them.

Nonverbal communication is more than just body language. It's how you walk onto a jobsite. It's how you sit—or where you choose to sit—at the subcontractor meeting. It's what you wear, your tone, your cadence. It's the behaviors and comments you allow, and the ones you don't. It's how you leave your work area for the crews or trades coming in behind you. It's how you show up. That "tone-setting" you hear people talk about at the start of a project? Nonverbal communication accounts for most of it—and it deeply influences how anything you say is received.

So much of our nonverbal communication is lost over emails, text messages, or even video calls. That's why it's so important that we invest energy into in-person interactions in addition to, or in lieu of, only electronic modes of communication. In-person interaction is what keeps our human element pipeline strong.

There are four main stages which occur during communication.

Stage 1: Motivation. This is our internal pipeline, mindset filtering, energy gauging, and feeling concoction process we covered in Chapter 5. Something is motivating you to want to communicate. You are looking for some type of action or result—outcome—from your energy and effort to communicate. It can be to defend your energy or fun level from an offender, or it could be to reach out and build a relationship with someone who has the potential to add to your fun factor. Either way, something is motivating you to communicate.

Stage 2: Action. This is when you attempt to communicate something to reach your desired outcome. This may be verbally, nonverbally, yelling, calmly, over email, over the phone, etc.

Stage 3: Listen. We are taking in cues—nonverbal and verbal—to see how our message has landed with the other person and if we are anywhere close to hitting our targeted outcome. Do they agree, do they want to fight, did they not understand one thing you said? This provides us with feedback on how to adjust our message to try and hit our intended outcome.

Stage 4: Respond. Adjust our message, language, tone, stance, whatever—to try to scope in and hit the target outcome of our communication.

If what you need to communicate holds a lot of weight and emotion, slow down and thoroughly inspect these different phases and map out the conversation. Identify what it is you need to convey to help you pave or maintain the road to fun and fulfillment in the industry.

Don't assume other people know how you feel, know what you are experiencing, or even know where you are trying to go and why. If you want it, you need to communicate it. This takes practice and some tact, but if you can't communicate it, then you will have to spend the energy to dig around the people standing in your path in lieu of simply asking them to either join you or kindly move out of the way.

Common Barriers and How to Overcome Them

While the idea of making construction fun again is appealing, many challenges might stand in the way. Below are common objections or obstacles, along with strategies to navigate them.

Objection: "My job is just too stressful to be fun."

Solution: Try reframing stress as a sign that you care about your work—it means you're engaged and invested. Rather than aiming to eliminate stress entirely, focus on managing it in healthy ways: take short breaks, delegate tasks when possible, and concentrate on what's within your control. Think of stress as a dial. Aim to stay between levels three and eight out of ten. If you hit nine or ten, use rest and refuel strategies to avoid burnout. If you're at zero to two, you might not be challenged enough, so try recharge tactics to prevent rust-out.

Objection: "I work with negative people."

Solution: Set boundaries and choose how you engage. Spend more time with those who uplift you, and limit interactions with energy leeches when you can. Lead by example—positive energy is contagious. Be the change you want to see.

Objection: "I don't have time for fun—there's too much work."

Solution: Fun doesn't require more time—it's about being intentional. Small shifts, such as injecting humor, celebrating small wins, or improving team dynamics, make a big difference.

Objection: "Leadership doesn't support this."

Solution: Change starts from the ground-up. While leadership plays a role, grassroots efforts—like fostering camaraderie within teams and demonstrating the benefits of engagement—can create ripple effects over time. Additionally, sometimes a company or industry sector's culture doesn't fit your personality, skills, passions, and strengths. You have to give it an honest try, but don't try to force the square peg into a round hole. This requires self-awareness of who you are and what you need to thrive. This also requires a little research by asking around, reading up on, and understanding the different sectors, projects, and roles, and their pros and cons.

Objection: "I'm burned out and don't know where to start."

Solution: Start small. Focus on micro-adjustments in your mindset, daily habits, and work relationships—small shifts make a big difference over time. If your stress levels are in the red zone, it's a sign to pause. Give yourself space to rest and refuel, so you can regain energy and clarity before making bigger changes.

You Got This

That's it! We now know the source of our leak: we've spent a bit too much time focusing only on efficiency, performance, and the latest tools, meaning that we've neglected the most important part of our industry—our people. Realizing this means we can now finally start the repair work to bring these elements back in balance and eventually effectively replace the drywall and give it a fresh coat of paint.

Don't forget who you are and all that you do. No one will do the hard work for you, so you must be intentional about rediscovering or finding and defining what fun is to you. Once you know this, plant it deep down in your gut and your heart, and allow it to be the foundation in how you show up every day, and connect and interact with others. There will be those who don't see it, and there will be those who never will. You have to be strong, and know down to your core that what you do is critical, impactful, and important.

You can't rely on others to provide this praise, motivation, or encouragement. You cannot rely on others to make the industry "fun again." You have to take ownership and responsibility to define and decide what fun is to you, and create it. Once you do, it will impact not only yourself, but others who may be looking for this spark as well.

Yes, the structures we create are phenomenal and fantastic, but it's memories that contribute to the true legacies we leave: the people we worked with, the struggles we overcame together, the laughter and anger and pride that was built into these structures. That is our opportunity to leave a legacy. It isn't the building or structure, it is everything that went into building that structure—the blood, sweat, tears, breath, life, history, and struggle that went into building that structure is what creates and leaves a legacy.

We don't point at a building and say, "It was built." We say, "I built that." The structure represents all the hard work it took to be a part of its existence. "I built that" means so much more. The experience and hard work it took to get there. It includes the relationships built, the losses, the gains, the everything.

To make construction fun again, *we* have to *make* it fun again. No one else will do it for us. Simply knowing just how amazing, gritty, and determined those of us in the industry are, I know you can do it. I believe in you. I appreciate you. You got this!

Afterword

EVERYONE IN CONSTRUCTION matters, so it matters that we look after everyone who contributes to our work and lives. The collective effort, sweat, humor, setbacks, and victories are there in every brick laid, every stick of pipe installed, and every weld.

Making construction fun again is a big task because it asks all of us to lead the change. But we don't move unless we know the pain of staying still. Right now, that pain is telling us that something's wrong.

We can see the stress and communication breakdowns; we can see how we put pressure on our trades to do it faster, cheaper, better, for the sake of the project—and how it sets up impossible expectations.

We've seen it in our leaders, the high-performing people who look like they've given up or are grinding their way to retirement or exiting the industry. We self-medicate to numb ourselves to the stress, while doing nothing to address its source. People take their lives and leave behind partners, kids, and teammates. We know friends in their fifties who have heart attacks because of the stress. I knew a general superintendent who was always so positive and dug very deep to keep up appearances; he died within a year of retiring. You could tell he'd given everything.

It doesn't have to be that way. When you come to learn that you're in control of it, it can be a big realization. We have the strength and the power to address the cause, not just the symptoms.

I've had an incredible career in construction thus far, working with exceptional people on countless projects. I have learned to see and create the fun in our industry, but, as you've seen, it wasn't always this way, and it hasn't been a walk in the park.

When I was putting the finishing touches on this book, I came across an old journal of mine from that particularly bad period I've mentioned, when I fell out of love with the industry in around 2011. I was surprised to see how I'd recorded how I was feeling at that time.

I was at a transition point, working as a project engineer and training to be a PM. The recession had seen a big chunk of my colleagues and peers laid off, and I had to take on new responsibilities far faster than I'd wanted to. This was the project where we'd taken over the build and remodeling of two schools over two summers, and I was to come in for the second summer as a manager.

> JULY 17, 2011
> So I bought this journal hoping to get me back on track with life. I feel so stressed about work. I have two schools that I am managing that are scheduled to be complete in two weeks. I am terrified that they will not open in time. I have lost sleep over the past three months, I work until about 6:30–7 p.m. every night and am extremely unhappy. It has been so stressful that I have seriously been considering quitting.

I HATE quitting. But this has been the most difficult and overloaded that I have ever been at [this company] and in my life.

I have been trying to calm myself and reevaluate my position but find my mind going a million miles per hour. That is why I bought this journal. In hopes to slow my mind enough to evaluate my thoughts. I figure that if I can remember the true reason I got into the construction industry I may be able to help with my stress level and cope with difficult, trying times.

I have also determined that I need to establish hobbies again. All my extracurricular activities are related to construction, which lately feel just as demanding as my job. So, on the next page, I will list reasons I got into the construction industry and hobbies I would like to try. I also want to go to Ireland. That is the main reason I chose this journal after starting and closely evaluating about twenty-five of them. This will be my motivation to get to Ireland!!

Construction is what I know and love. Even though it is very difficult right now, it is worth it to stick it out, because I love it. I love being associated with it. I love knowing that I have that knowledge base.

Why I got into construction:

- I want a sense of accomplishment. I like seeing the buildings once they are complete.
- I wanted to be the translator between the contractor and owner. I started from the bottom and have the knowledge to do so.
- I wanted to prove people wrong in the sense that they would underestimate me as a dumb blonde. But I actually know what I am doing.

- I wanted to have the comfort and security of knowing that I could build something if I needed to.

Hobbies:

- Skateboarding
- Saxophone
- Gardening
- Build miniature models

You must and will stay calm in stressful situations. All you can do is your best. All you have are your values. Which, at the end, will supersede any situation. There is absolutely no reason to get stressed out about work. If something couldn't get done, then it was unobtainable because you put your all into everything you do. If it ends in you getting fired, then there was nothing you could do to prevent it anyway.

There are a few more entries, but not many. What they tell me, in hindsight, was that during the most difficult times in my life, I felt that I needed to make some sense of what was going on by putting it down on paper.

This one appears a few days later, on July 20, 2011:

One week and two days left of the hardest project I have ever had. I have been listening to calming music to and from work to help me relax and it's nice to daydream about the mountains like I used to in college. I find myself daydreaming of happy places. Like my mind is forcing me to, to equal out the bad and good.

Another from September 8, 2011:

So I am finally done with the rush of the most difficult projects of my life to date. I am still finishing things up and it is still challenging, but I am finally on the downward slope.

The good that has come out of this major challenge is that I have been forced to re-evaluate my life values. On August 19th (my wedding anniversary), I found myself in the hospital at 12:00 at night. They deduced it to food poisoning, but considering I still feel traces in my stomach when I find I am worried about something, I believe it was due to stress. I needed to re-evaluate what I was doing, and how I was doing it.

I have started reading *7 Habits of Highly Effective People* for the second time, and it made me realize that my stress was created only by my perception of events.

I was creating this life-or-death, fight-or-flight response to situations, that are important, but not so important that it compromises my health and time with the person I love the most, my husband, and those I care about most, my family.

I need to write down my mission statement/values to remind myself of my life goals and purpose. I have no idea how to even start that but I will at least try.

Another breath of fresh air is that I am responsible. It seems like it would be a burden to say that, but it is the most refreshing thing to realize. I can't blame circumstances, situations, or people when I am upset, happy, unhappy, etc. I choose what situations make me feel that way, and it is my perception that makes me responsible. No one can take my happiness from me unless I give it to them.

This was a turning point for me, because I seemed to start directing my attention to changing my mindset. The next entry doesn't appear for almost eighteen months:

Well, I am still employed, and had a few great projects last year. I have received a raise and bonus every year. I am

> still figuring things out, but don't want to write about work today. I am seventeen weeks pregnant as of today. I am so happy and excited. All I want to do is go to the doctor every day to hear the heartbeat to make it seem more real...

Those school projects were a big part of my life. We finished on time, but at a cost. It was not a financial win, and during the Great Recession we needed wins.

Years later, I went to an event and happened to sit down next to the person who was my executive at the time. We talked about challenging times in our careers, and the benefits and learning that we took away. My former executive reminded me that the client was extremely happy, and the architects and subcontractors all felt supported during the process. In fact, I was called years later by the owners' representative for that client, who offered me a job to fill his role upon his retirement. Yes, we lost money on it, but the long-term result was far better: we'd nurtured that human pipeline during a difficult period, and the people in our industry were better for it. Yes, there were people from that era who didn't like me (as my snarky emails will attest!), but there was a lot learned, too. It taught me that I was ready to be a PM.

Compare that with my stressed journal entries at the time, and you can see how only looking at what's immediately in front of you can really change your outlook.

As I've said before, when you get it right, the change is incredible. It's what inspired me to step away from being a PM and director of training (both of which I truly enjoyed) to get the message of change out further. I just knew that a program needed to be built—one that gets to our people, gives them the tools they need, and helps prevent those disastrous outcomes of construction not being fun again. I started

Well Works and built a Leadership Development Program, which helps equip frontline construction leaders with tools and information to efficiently and effectively communicate with—and lead—our project teams in our demanding and fast-paced industry.

It took me a year to build the program, and I quality-tested it with a small group of industry experts of varying ages and levels of experience. The feedback was incredible. They knew about the human pipeline instinctively—we often do—but now they could understand what was happening and why. They could be strategic and intentional about improving their conditions.

I've worked with many organizations since then, and the response has always been the same: "Thank you for helping us care for our people." That's a huge reward that keeps me going. When you get the chance to change our industry for the better and see that it's had a material impact on someone who's still in the industry ten years later and loving what they do and who they work with, that's proof that making construction fun is possible.

Be the energy our industry needs. Be the hand to help someone out of the pit.

Behaviors are contagious. It will take time, but we know that. No project gets completed in a day, but a part of it does. So every day is another chance to shine a light into places that have been dark for too long, and, with time, we'll have re-shaped our industry. We'll have done some good work along the way.

Appendices

Appendix 1
Human Element Pipeline Section Details

In Chapter 5, we introduced the human element pipeline—our internal system for taking in, filtering, and responding to information on the job. Think of this appendix like diving into a specific detail sheet in a set of construction plans. The big picture may already be laid out in the floor plans, but it's in the details—those sections, elevations, and callouts—where you understand how everything actually comes together.

This section gives you a closer look at the inlet gate (our senses), as well as the heart-brain and gut-brain networks that often operate beneath our awareness. Just like a plan set shows both the structure and systems of a building, this appendix maps out the structure and systems of our internal processing—how we take in information, how it affects us, and how it shapes what we build together.

Inlet Gate Detail: Our Senses

Information is constantly absorbed through our senses, which serve as an inlet that channels data into our processing system. Consider these senses—the ones we use to hear, see, feel, smell, taste, and perceive—as our inlet gate.

We may not be able to control everything that floods in through this gate, but we can try to filter it by attaching a "trash grate," adjusting the direction of the flow and throttling how much information enters at any given time. For example, while we sleep the inflow is reduced to a mere trickle, yet the moment we wake the torrent of data resumes, and the process of filtering and synthesizing these inputs into our mood and behavior begins immediately.

Consider our work boot scenario from Chapter 5: the moment we drove onto that jobsite, we were inundated with environmental information. What did you notice? Were materials scattered like clothes and toys in a ten-year-old's bedroom, or were they segregated, neatly stacked, sorted, and almost suspiciously clean? Where were others parked? How were they dressed, how did they interact with one another, and handle their tools? What did the air smell like—mud, exhaust, sawdust, burning metal from welds, leaking gas, or even body odor from someone who skipped their morning shower? What sounds filled the space—arguing, laughter, chatting, drilling, sparks from welding, backup alarms, or was it eerily quiet? What overall impression were you left with?

Heart-Brain and Gut-Brain

Our bodies gather information not only through our traditional senses but also through other neural networks, and we process this data with more than just our conscious mind. In the 1990s, scientists discovered neurons—brain cells—in our hearts and guts. These clusters, now known as our heart-brain and gut-brain, also absorb and process inputs from the external environment, including electromagnetic waves.

Our heart houses the body's most powerful magnetic and electrical fields. The heart-brain stores memories and experiences and communicates them directly to our conscious mind, using a language of feelings and instincts rather than words. Additionally, the heart produces and secretes oxytocin, a crucial neurotransmitter for social interaction and connection. According to Rollin McCraty, Ph.D., Director of Research at the HeartMath Research Center: "The magnetic field produced by the heart is more than 100 times greater in strength than the field generated by the brain and can be detected up to 3 feet away from the body, in all directions."[33]

Similarly, our gut-brain operates much like our heart-brain. It can function independently from our conscious mind, thanks to its own reflexes and sensory mechanisms that absorb environmental information. As one source explains: "There are more neurons—100 million, to be precise—in our intestines than in our whole spinal cord. As with the heart brain, we cannot consider the gut brain as a mind, per se, but its influence on the upper brain is nonetheless very clear."[34] Moreover, most of our serotonin—a neurotransmitter that calms aggression, reduces impulsiveness, and enhances sociability—is produced in the gut.

Many of us can sense the mood of a trailer or the tone

---

33 R McCraty, *Science of the Heart, Volume 2: Exploring the Role of the Heart in Human Performance* (Boulder Creek: HeartMath Research Center, 2015), 38, https://www.heartmath.org/research/science-of-the-heart/.

34 Carlos Davidovich and Jennifer Elizabeth Brunton, *Five Brain Leadership: How Neuroscience Can Help You Master Your Instincts and Build Better Teams* (Vancouver: Page Two Press, 2023), 18, https://www.amazon.com/Five-Brain-Leadership-Neuroscience-Instincts-ebook/dp/B0BMHIR3LC/.

and pace of a project before a word is spoken. This intuitive ability stems from our past experiences as well as the combined filtering and processing performed by our heart-brain, gut-brain, and conscious mind. Together, they use neurotransmitters and hormones to generate meaning, which then influences our actions, reactions, and interactions within the environment. Consider again our work boot scenario: What did you sense when different people entered and either took their seats or remained standing during the meeting? What impression did you have when you first stepped into that trailer?

Putting the Details to Work

Just as you wouldn't build a structure based solely on a cover sheet or floor plan, we can't fully understand human performance and connection without looking into the sections and systems that run beneath the surface. The Human Element Pipeline might not be detailed in the spec book, but it is constantly influencing how we interact, interpret, and lead.

Use this understanding the same way you'd use a detailed section drawing: to anticipate how things fit, to identify where pressure points might exist, and to support alignment across the whole system. When you're walking onto the jobsite, into the trailer, or into a conversation, these are the specs that help you show up with greater clarity, purpose, and precision.

Appendix 2
Recharge: Fuel Up Worksheet

Before you can lead others with focus and energy, you need to know what fuels you. This isn't about kicking back and relaxing—it's about what actually recharges you.

Use this worksheet to identify what recharges your battery—the things that leave you feeling sharper, stronger, and more ready to show up for your team and your work. This isn't one-size-fits-all. It's your personal fuel map.

1 **People**
Who leaves you feeling more focused, motivated, or charged after a conversation or time spent together?

2 **Environments**
What spaces, locations, or surroundings tend to recharge you or boost your motivation?

3 **Activities**
What do you do that leaves you feeling more energized afterward—even if it's hard work?

4 **Moments in the Past When I Felt Most Energized**

Think of a time when you felt strong, focused, and driven. What was happening? Who was there? What were you doing?

____

____

____

5 **Fuel Drains Versus Fuel Gains**

Are there regular things in your routine that drain you? What could you replace them with to gain energy instead?

Drain: __________________________________
Replace with: ____________________________

Drain: __________________________________
Replace with: ____________________________

Drain: __________________________________
Replace with: ____________________________

Field Application Notes

Once you've filled this out, don't shove it in a drawer. Post it in the trailer, keep it in your glovebox, or wherever you'll see it often. These aren't just "nice-to-haves"—this is how you keep your systems running strong.

You're one of your crew's most important tools. Stay fueled so you can keep showing up solid, clear-headed, and ready to lead.

MENTAL HEALTH IN CONSTRUCTION

WHERE TO TURN

If the stresses of work—or life—feel too heavy, you're not alone. Mental health challenges and substance abuse are unfortunately common in our industry. You don't have to navigate this alone. Here are some places to start:

988 Suicide & Crisis Lifeline (US)

Call or text 988.

Construction Working Minds (US)

ConstructionWorkingMinds.org for industry-specific mental health resources.

National Helpline (SAMHSA)

1-800-662-4357 (US) offers confidential treatment referrals and information.

Man Therapy

ManTherapy.org uses humor, relatable language, and a "20-Point Head Inspection" quiz to help men—especially those in high-stress fields—tackle mental health challenges head-on.

Global Resources

If outside the US, visit findahelpline.com to search for local hotlines.

It's okay to need help. Reaching out is the first step toward rebuilding a solid foundation—both for you and the people who rely on you.

Glossary

A

Adrenaline: Our action and adventure neurotransmitter. Where cortisol is more the alarm, adrenaline is that release of energy into your system. That surge of energy can feel good, especially when you are working out or enjoying high-octane activities like motocross or skydiving. It can also feel a bit nerve-wracking, like when you are in a job interview or challenging an alpha for their spot at the top. Your pupils dilate to let more light in, your hearing sharpens, and the brain processes what you perceive more quickly. It creates this state of concentrated attention that can give you access to more information about your physical environment, and it gives you energy and focus in important or critical situations.

As-builts: A final set of drawings that represent all the changes and final placements of systems and materials throughout a project essential for future maintenance, renovations, or modifications. They provide a detailed record of the finished structure. The contractor marks up and posts any and all of these changes, and then delivers them to the owner's rep and facilities maintenance representative where they are promptly thrown in a closet with all the mandatory extra materials that are rarely used where they will eventually turn to dust, never to be seen or found again.

ASIs (architectural supplemental instructions): An official project document that provides clarified or additional information to the contracted drawings. Industry etiquette is that all changes should be clearly outlined, clouded, or delineated, and that the changes should not require additional cost or time.

B

Brainstem (a.k.a. our reptile brain): A region of our brain. The brainstem is buried and nestled under the cortex and the limbic system, right at the back. If you were to locate where your spinal cord connects with your head and skull, it's buried and protected deep in that area. This region is like the fire sprinkler and electrical and mechanical rooms all combined into one area. It regulates our survival systems like breathing, digesting food, and keeping our heart beating. It is involved in reactive, automatic, primal functions like freeze, fight, flight, fawn: instant reactions that help us increase our chances of survival in threatening situations. This is where our energy override switch resides and can ultimately shut off power to the main control system in the pre-frontal cortex. It is also where some of our survival neurotransmitters and hormones are injected and mixed into our pipeline contents.

C

Catty-corner (a.k.a. kitty-corner): On the diagonal corner. For instance, if you drew a square and your pencil is on the bottom-right corner, catty-corner from your pencil would be the top-left corner.

Cerebral cortex (a.k.a. cortex): The part of our head-brain that our old-style hard hat protects and covers. The pre-frontal cortex is obviously a part of this, but it is the larger mass that the majority of your skull covers and protects. This region makes maps of our life and experiences. It is the part of the brain that weighs situations logically and rationally and generates plans of action.

Cortisol: Our safety-seeking survival neurotransmitter. It is responsible for the feeling of fear or stress and ignites the freeze, fight, flight, fawn survival mechanisms. Usually, something we care about is at stake and motivates us to do something about it. Cortisol turns our stored sugar and fat into energy, which gives our bodies and brain the ability to use that energy efficiently and effectively to handle the situation and try to eliminate that fear or mitigate that stressful situation. In order to do this, it turns off anything that is not essential for handling that fear or stress, including digestion, reproduction, and growth, and dampens our immune system functions. By the way, dehydration increases cortisol levels, where hydration is associated with significant decreases in cortisol levels. Studies are currently testing whether cortisol has negative effects on perception, spatial ability, and memory. Not so great on our construction projects.

D

Dopamine: Our reward-seeking neurotransmitter. This is responsible for the feelings of motivation, drive, and anticipation. It is beneficial but highly addictive and overpowers other beneficial neurotransmitters if we don't engage impulse control every so often. We feel this when we hear dings or feel vibrations from our phones or watches when we receive a notification. The smell of french fries or donuts triggers this. It is the dangling carrot, so to say. It's the anticipation and seeking of a reward—the reward being "feel good" responses like fulfillment, accomplishment, respect, security, comfort, and status that accompany this neurotransmitter.

E

Endorphins: Our comfort and feel-good neurotransmitters.

G

Gut-brain: The gut-brain is located exactly where it sounds, in our digestive system. We all have had that gut instinct or a "gut feeling" before. This is our gut communicating in its own language and using that same vagus nerve conduit our heart-brain uses to send messages to our logical brain. Most of our serotonin, a neurotransmitter that helps calm our aggression and impulsiveness and make us more sociable, is produced in our gut. According to Carlos Davidovich and Jennifer Elizabeth Brunton, "There are more neurons—100 million, to be precise—in our intestines than in our whole spinal cord. As with the heart brain, we cannot consider the gut brain as a mind, per se, but its influence on the upper brain is nonetheless very clear."[35] The gut-brain can control its behavior independently from the brain because it has its own reflexes and senses to absorb information from the surrounding environment.

---

35 Davidovich and Brunton, *Five Brain Leadership*, 18.

Guys: A term used to address a group of people (male or female), as in, "Hey, you guys." Most frequently used (but not exclusively) by unassuming Millennials, picked up from movies and entertainment in the 1990s and used for groups they like, appreciate, and respect. Used to be inclusive rather than intended to be derogatory, offensive, or exclusive.

H

Happiness: The word we use for the feeling produced when four different neurochemicals (neurotransmitters) are activated: dopamine, endorphin, oxytocin, and serotonin.[36]

Heart-brain: This region of our brain is located on the left side of our chest, protected by our ribcage. It's that area behind the pencil-and-notepad pocket of your safety vest. Yes, it is indeed your actual heart, where in the 1990s scientists discovered neurons. This region stores some of our memory and experience filters. When the contents of our pipeline get to this region, it is filtered through these sieves, injected with any applicable memories from experiences, and transmitted as lessons and messages into the output pipeline, directly to our brainstem and limbic regions. It does this using our vagus nerve located in our spinal cord like electrical wire in a conduit. These heart-brain messages aren't transmitted in words but in the form of feelings and instincts, and they impact our decisions, behaviors, and motivation. Our heart also manufactures and secretes oxytocin, an important neurotransmitter for interacting and connecting with other humans.

36 Breuning, *Habits of a Healthy Brain*, 13.

I

Inlet gate: A structure designed to control the flow of liquid entering a pipe, acting as a controlled entry point.

K

"Keep emotions out of it": For the sake of our entire industry, allow me to clarify what this statement means. It means that you need to understand the value and importance of your emotions enough to consider them and further control or manage them in your interactions. It's a reminder to not to take things personally, as whatever is being discussed or worked on is for a greater, larger team, company, or project outcome.

L

Limbic system (a.k.a. our mammal brain): A region of our head-brain. If you were to touch your temples and move your fingers back to just above your ears, the limbic system is buried somewhere deep in that area. Much of it is covered by your cortex. This part of the brain is also known as our "mammal brain" because it operates similarly to the brains of other mammals, such as dogs, cats, elephants, dolphins, etc. This is where our emotions and GAS (Give-a-Sh*t) meter reside. It's also where our motivation comes from and is involved in memory. This region holds a large amount of mindset lenses as well. This is where much of our content within the pipelines is mixed and infused with some of the neurotransmitters and hormones. This mixture initiates and creates those feelings of fulfillment and enjoyment—important for connecting with other humans.

M

Machiavellian: This describes individuals with a strong focus on self-interest over the team's or project's well-being. They will prioritize personal gain, influence, or power, sometimes at the expense of trust, collaboration, and ethical standards. They come off as very technical- and process-savvy, hitting budgets and schedule deadlines, masking the wake they leave in their path to those in leadership. But with their kiss-up and kick-down mentality, they are usually toxic to peers or anyone working underneath them.

Masonary: This isn't a word but is frequently used and is the way our industry pronounces "masonry," for some reason. No matter how hard you resist or try not to, the longer you work in our industry and interact with our fantastic field folks, the more likely you will start saying "masonary," too, whether you like it or not.

Mindset: How you choose to see specific people, situations, environments, and industries (i.e., your worldview). This includes the lenses you get to create, switch out, and adjust, which create beliefs about your reality and which greatly impact your behavior, actions, and interactions.

Mindset, Fixed/Closed: Those with fixed/closed mindsets believe that success comes from one's innate ability and thus avoid challenges for fear of failing. They give up easily because they do not believe that effort pays off.[37]

---

37 Sousa, *How the Brain Learns*, 60.

Mindset, Growth/Open: Those with growth/open mindsets believe that their success comes from their efforts and persistence. They rely more on hard work and resilience than on ability for their achievement.[38]

N

Neurochemicals: Usually categorized into two groups: neurotransmitters and hormones. The ones that stay in the brain are called neurotransmitters, and the ones that cross the blood-brain barrier and circulate in the bloodstream are called hormones.[39] Although, it seems that scientists may have recently discovered neurotransmitters in the heart, which indicates that some neurotransmitters are therefore capable of crossing the blood-brain barrier.

Neurotransmitter: Any of several chemical substances, such as epinephrine or acetylcholine, that transmit nerve impulses across a synapse to a postsynaptic element, such as a nerve, muscle, or gland.[40]

Noradrenaline: A neurotransmitter that is at an "optimal level when you feel slightly overchallenged; it leads to a 'this is tricky, but I think I can handle it' feeling. It is also released when you push yourself to perform a difficult task better, faster, or with fewer resources."[41]

---

38 Sousa, *How the Brain Learns*, 60.

39 Breuning, *Habits of a Healthy Brain*, 24.

40 Dictionary.com, s.v. "Neurotransmitter," accessed October 8, 2024, https://www.dictionary.com/browse/neurotransmitter.

41 Friederike Fabritius and Hans W. Hagemann, *The Leading Brain: Powerful Science-Based Strategies for Achieving Peak Performance* (New York: TarcherPerigee, 2017), 10.

O

One-line: A simplified or highlighted view of electrical or piping systems. It's as if you had x-ray goggles and could look into a building from an arial view to see these different systems. It shows the lines, risers, outlets, inlets, etc.

Oxytocin: Our human connection-seeking neurotransmitter. Responsible for the feeling of trust, it is ignited when interacting with others. It is longer-lasting than dopamine and actually reduces the addictive nature of dopamine. It dampens our fear (freeze, fight, flight, or fawn) response and enhances our communication. Some call this our "tend and befriend" neurotransmitter, but it has its dark side, too. "Humans divide others into at least three categories: members of liked groups, members of disliked groups, and strangers whose group affiliations are unknown."[42] Oxytocin can promote hostility toward members of disliked groups.

P

Personal protective equipment (PPE): Clothing and accessories that help protect yourself from different elements. This can include safety glasses, safety vests, steel-toed boots, gloves, hard hats or helmets, etc.

---

42 Matthew D. Lieberman, *Social: Why Our Brains Are Wired to Connect* (Oxford: Oxford University Press, 2013), https://www.amazon.com/Social-Why-Brains-Wired-Connect-ebook/dp/B00CCPII14/.

Pre-frontal cortex (a.k.a. main control center): A region of our head–brain. If you were to put on your hard hat (not the new and improved helmet style, but the old-style hard hat, which is still the current style for many of those in our industry), notice that area where your sweatband is. Directly behind this area is your pre-frontal cortex. It is the main control center of all the different regions of the brain. It houses our executive functions, like working memory, planning, problem-solving, impulse control, critical thinking, and emotion regulation. This region requires a lot of energy to operate and houses some of your mindset lenses.

R

Request for information (RFI): A tool used to clarify the intent of what is wanted or expected (by the owner or design team) in order to build the building correctly and as intended. It is typically a clarification of, or a request for clarification of, information that may not be clear or may even be missing from the construction documents. These are an important communication tool used throughout the entire project. Although we sometimes get verbal direction from an owner or an architect, we must follow up with a written RFI to ensure the information is clearly communicated to all of our other team members working on a project (subcontractors, suppliers, superintendent, principals, etc.).

S

Safety data sheets (SDSs): Previously known as "material safety data sheets" (MSDSs), these documents are like ingredient and nutrient labels on steroids. They include the physical properties, health effects, storage and disposal, safe handling, recommended PPE, and spill and leak procedures for different materials and chemicals.

Serotonin: Our status-seeking neurotransmitter. It is responsible for the feelings of pride, accountability, and respect. It regulates our emotions and moods and enhances our perception, intuition, and self-control. It raises our status within a pack or group and is activated anytime we choose to help others.

Submittals: Submittals are documented forms of communication used to verify that the proposed materials to be used or installed on a project match the intent of the architect's/engineer's/owner's design requirements. This documentation process is typically required before materials can even be ordered, so it has to be done early in the project to make sure materials are verified, ordered, delivered, and on site when they are ready to be installed.

W

Work-life integration: Who you are at home and in your personal life adds to who you are and how you show up at work. And what you do at work—whether it's learning skills, working hard, or leading a crew—shapes who you are when you go home. It's all connected. Being a good parent, spouse, or friend can make you a stronger, more focused worker. And doing your best on the job builds confidence and pride that carries over into your personal life. Work and life aren't separate—they work together to make you better in both places.

Acknowledgments

THERE ARE COUNTLESS people who, in ways both big and small, have contributed to the making of this book. Some I've had the privilege to work with directly, others I've had engaging and energizing conversations with. Whether through mentorship, collaboration, inspiration, or by challenging the status quo, each has left an indelible mark on my thinking and, by extension, on these pages.

While this book reflects my personal experiences, insights, and conclusions—along with the inevitable imperfections that come with that—I want to extend my deepest gratitude to those who generously shared their time, knowledge, and perspectives. Their contributions have enriched this work immeasurably. However, it's important to respectfully acknowledge that their involvement does not necessarily imply agreement with all the ideas, opinions, or conclusions expressed in this book. Any errors, oversights, or controversial takes are entirely my own, and I alone bear responsibility for them.

I would especially like to thank a handful of individuals who offered direct support, feedback, and guidance. Some of these contributors have chosen to remain anonymous for personal or professional reasons, and I honor and respect their wishes completely. To those who gave their insights quietly, behind the scenes, your impact is deeply appreciated. Thank you—from the bottom of my heart—for your generosity, your energy, and your dedication to supporting this extraordinary industry.

For those who have generously and proudly allowed their names to be included, I am honored to acknowledge their contributions here:

Alicia Ready
Training and Development Director;
Commercial General Contractor
Over 18 years of experience in commercial general contracting, transitioning from project management to people development.

Brad Motherway
President, Gregory Electric
Over twenty-one years of experience in the construction industry. Began his career at eighteen with no prior experience and rose to company president through dedication to the trade and a deep commitment to the values of hard work and growth.

Cal Beyer, Beyer, CWP (Certified Wellness Practitioner)
Senior Director of SAFE Workplaces, SAFE Project
Over thirty-five years of experience in human capital risk management, with a focus on the construction and manufacturing sectors since 1996.

Christofer M. Harper, PhD
Associate Professor & Chair of Heavy Construction Management, Colorado State University
Thirteen years of experience in plumbing, general contracting, and owner representation. Now a researcher and educator in workforce development, staffing, and construction technology, contributing to over twenty-five industry research projects.

David Johansson, MA
Owner and Founder, Triple C Coaching
Thirty-eight years of experience in electrical construction. Known for major projects such as Denver International Airport and for his focus on leadership and decision-making through the lens of neuroscience.

Dylan Niadna
Site Superintendent
Thirteen years of experience managing nearly $1 billion in new construction and renovation projects across Washington and Michigan. Holds a BS in Construction Management from Ferris State University.

John Spight
Division Manager; National Commercial General Contractor
Over eighteen years of experience in the construction industry, leading major commercial projects and teams across the US.

Katie Barnes, AIA
Principal & Managing Partner, Barker Rinker Seacat Architecture
Over twenty-four years of experience focused on community-based public architecture, with a passion for improving design and construction processes to better serve communities.

Matt Powell

Co-Owner, Well Works; Operations Management Consultant and Managing Partner, Technician Services Business

Over twenty-four years of experience in the construction industry, with a strong foundation in subcontractor operations. Specializes in data center project delivery and operations management.

Michael W. Millsapps, CFM

Executive Director, Facilities Management

Thirty-eight years of experience leading strategic and operational facilities programs in institutional and commercial environments.

Pete Parizo

Workforce Development Manager; Heavy Civil Contractor

Over twenty-six years of experience in the construction industry, with a focus on workforce development in infrastructure and heavy civil sectors.

Ron Nussbaum

Founder & CEO, BuilderComs; Host, Construction Champions Podcast; US Marine Veteran

Over ten years of experience in construction. A passionate advocate for improving field communication through technology and leadership.

Tal Clark

General Manager; Mechanical Contractor

Forty years of experience in mechanical construction, involved with projects ranging from small-scale to over $80 million for commercial and institutional clients.

Ted Johnson

President & CEO, ECI Site Construction Management, Inc.
Thirty years of experience in the construction industry. Graduate of Colorado State University's Construction Management program, Class of 1999.

With Special Thanks

I would also like to personally thank the following individuals who played a critical role in helping me throughout the writing and publishing process. Your guidance, encouragement, and generous contributions, whether through editing, coaching, or behind-the-scenes support, meant more than I can express.

- Matt Powell
- Alicia Ready
- Sharon Ulrich
- Dave Taylor
- Cameron McKague

MICHELLE BADGER PHOTOGRAPHY

About the Author

AMY POWELL is the founder and CEO of Well Works, a company devoted to developing and delivering essential "people" skills programs, workshops, training, coaching, and consulting for professionals working in or partnering with the construction industry.

With over two decades of construction experience, Amy has served as a project manager at a wonderful mid-sized commercial general contractor. Her husband (and Well Works co-owner), Matt Powell, is also a construction professional with more than twenty-five years in the industry in roles ranging from laborer to executive management. Much of his experience has been in the subcontractor realm. Over the past eighteen years of their marriage, they've often found themselves chatting about the construction world after work—conversations that were anything but boring. Little did they realize that those late-night talks were quietly shaping the unique skills, insights, and perspectives that ultimately inspired them to launch Well Works and write this book.

As many in the field know, a project manager's primary mission is to support superintendents and field leaders by ensuring they have the contracts, materials, budgets, design-team decisions, and owner approvals needed to keep projects moving. Amy loved every moment of it!

In 2013, shortly after her first child was born, she sought a healthier work-life balance (integration) and transitioned into a director of training role (while still working part-time as a project manager). That change opened her eyes to an important reality: a team's success depends not just on solid management and technical expertise, but also on vital soft skills—like leadership, emotional intelligence, and communication. She discovered that practical, hands-on programs teaching these skills were hard to find, especially for "boots on the ground" team members.

A few years later, tired of explaining the construction industry's unique needs to outsiders, Amy decided to roll up her sleeves. She earned a Master's in Adult Education and Training and founded Well Works. Today, her commitment to supporting field teams and the broader construction community remains unwavering. In addition to building and developing programs, services, consulting, and coaching with Well Works, Amy is actively involved in multiple sectors of the industry—serving on boards, participating in research projects, volunteering with industry-focused initiatives and events, speaking at conferences and webinars, and even helping create an annual event called Construction Con, which offers high school students a hands-on look at the many career paths in the construction field.

When she's not immersed in construction, Amy enjoys overlanding and camping with her family, reading, practicing mixed martial arts, and exploring mountain trails. From her first seat in "Construction 101" at Colorado State University's Construction Management program to helping teams develop the soft skills essential for success, her journey has been defined by the camaraderie, grit, and determination she has encountered in this industry. She wouldn't have it any other way.

www.ingramcontent.com/pod-product-compliance
Lightning Source LLC
LaVergne TN
LVHW100304210926
848480LV00003B/1016
* 9 7 9 8 9 9 2 5 2 1 4 0 5 *